Introduction.

Thank you for picking up this book!

Recently, DX (digital transformation), IoT (Internet of Things), and AI (artificial intelligence) have become a hot topic in the manufacturing industry. Then, when you want to specifically innovate or improve your business using DX, IoT, and AI, where should you start? I consider it a "manufacturing site". This is because innovations and improvements in the "manufacturing site" are most directly linked to sales and profits.

The manufacturing industry is a "company that makes things." There is also a lot of manual work depending on the product to be manufactured. However, considering the era of declining birthrate and aging population in the future, automation of manufacturing using machines and robots will be promoted. PLC (programmable logic controllers) that issue various operation commands to machines and robots and touch panels that enable the operation and monitoring of machines and robots are indispensable for automation technology in various manufacturing industries.

This manual explains how to create touch panel screens used in the manufacturing industry as carefully and as clearly as possible. We hope that by gaining knowledge of creating screens for touch panels, which are the mainstay products of factory automation (FA), you will be able to improve the efficiency and workability of your worksites, and reform and improve your worksites by utilizing DX, IoT, and AI, even if only a little.

"I'm going to use it for work, so I want to learn about touch panel screen creation!" "Creating touch screen screens for job sites and equipment There is no book that describes it in detail..." "I've read books on touch screens, but I don't feel like I can create a screen just from classroom learning..." "I have been studying through e-learning and other means, but the content is boring, and I have fallen behind..."

We have created the following three concepts so that these people can enjoy learning until the end of the course. (The screen creation software is a free trial version.)

Specialized contents for screen creation.
⇒ Deepen understanding of touch panel screen creation

Explanation using familiar objects as examples.
⇒Easy to visualize in your mind

Easy to visualize in your mind.
⇒Practice questions (15 questions in total) to actually create screens.

*This is a familiar example of a touch panel that is not actually controlled by a touch panel to make the explanation easier to understand.

This manual also introduces the PLC programs necessary for desk top simulation after the touch panel screen is created. I have been involved in PLC programming and touch panel screen creation for many years, and I used to think that PLC programming and touch panel screen creation could only be acquired through on-the-job experience.

However, as a result, I have seen many people get fed up and quit in the early stages... I decided to create this book because I want many people who are thinking of learning touch panel production to think, "Touch panel production may be surprisingly fun. In the coming era of declining birthrate, the number of people who can do niche jobs such as PLC programming and touch panel production will be decreasing rapidly. Conversely, however, it is precisely because they are niche jobs that they can become "valuable human resources with special skills.

We hope this book will help you to become an excellent touch panel screen creation engineer who will be responsible for the factories and facilities of the future.

Chapter1: Overview of Touch Panels

1-1 What is a touch panel?

Simply put, a touch panel is "**a screen that monitors and operates the status of amachine**". Touch panels are mainly used in connection with a PLC (programmable logic controller), a controller that controls equipment. The photo below shows atouch panelmanufactured by Mitsubishi Electric.

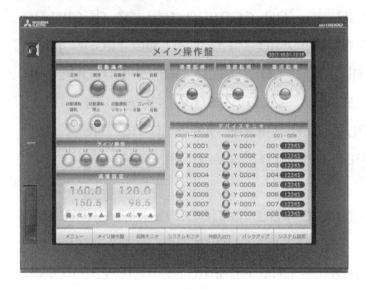

Various controls are possible by sending signals and numerical values from the touch panel to the PLC, which in turn sends signals and numerical values to robots and machines. It will also be possible to receive signals and numerical values from the PLC and display them on a touch panel to determine the state of the robot or machine. Thus, touch panels play a very large role in various fields.

1-2 Familiar examples of using touch panels.

For example, a familiar example of the use of touch panels is conveyor-belt sushi.

The main roleof the touch panel in kaiten sushi is,

Order sushi. ⇒Notify the kitchen of your order.

Voice alerts when the sushi you ordered is about to arrive.
⇒Prevent forgetting to pick up ordered sushi.

Press the checkout button to pay.
⇒Tell the clerk that you are going to pay the bill.

Switch display to foreign languages (English, Chinese, Korean, etc.)
⇒Orders can be placed by foreign customers.

and others.

1-3 Most common applications for touch panels.

The most common application of touch panels is in factory operation and monitor screens, where touch panels are used to operate and monitor various equipment and machines in factories. Although PLCs control robots, machines, conveyors (machines that transport objects), etc., touch panels are necessary for applications such as grasping the status of equipment and machines and switching to manual operation for trouble recovery. Factory automation (FA) = factory automation can be promoted by using PLCs and touch panels.

Factory Automation

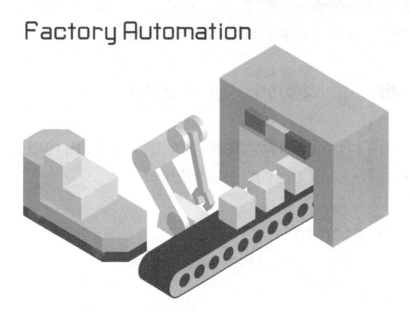

Also, factories are located all over the world, so if you can create PLC programs and touch panel screens, you will be able towork in factories overseas. PLC programming and touch panel screen creation is a niche field compared to C and JAVA like IT engineers, but because of this, there are few people who have mastered PLC programming and touch panel screen creation, making their skills rare and valuable. I have seen various factories and it seems that few factory workers can modify, add, or change PLC programs or touch panel screens. Learning PLC programming and touch panel screen creation will increase the scope of equipment improvement, so please takethis opportunity to learn it!

1-4 Major TouchPanel Manufacturers for Factories in Japan

These are touch panel manufacturers for factories in Japan, but there are various companies such as Panasonic, Fuji Electric (Hatsuko Electric), Mitsubishi Electric, Keyence, and Schneider Electric (Pro-face).

Touch panels in factories are often used by the same manufacturer as PLCs. (Another manufacturer can also be used.)

The major PLC manufacturers in Japan are Mitsubishi Electric, Keyence, and OMRON, and thesethree companies are said to account for about 90% of the total market share, so the major touch panel manufacturers arealso Mitsubishi Electric, Keyence, and OMRON. Of these, Mitsubishi Electric has a particularly large share of the market. If you have not yet decided which touch panel manufactureryou would like to study, we recommend that you start with Mitsubishi Electric's touch panel screen creation.

1-5 Reasons for using touch panels in factories.

The main reasons for using touch panels in factories include.

1. easy to add buttons and lamps
If you want to add buttons or lamps, hardware modification is necessary if there is no touch panel, but if there is a touch panel, software modification can be used. As a result, it leads to cost reduction.

2. monitoroperation status
Equipment operatingstatus, production count, defect count, and operation status of each machine can be checked on the touch panel. Also, if the data is in the PLC, the monitor status can be added to the screen.

3. anomalies and history of anomalies can be checked on the monitor
The touch panel can display the details of the error and the method of correcting the error, which leads to early recovery from the error. In addition, the system can keep a history of abnormalities, making it possible to make improvements to facilities based on the history of abnormalities.

4. high and low temperature, vibration, shock, etc.
There are also models with enhanced environmental resistance, and they can be used in a variety of environments, which is another reason why they are widely used infactories.

1-1 to 1-5 Summary

A touch panel is a screen that monitors and operates the status of amachine.

The most common application for touch panels is factory operation screens, where touch panels are used to operate and monitor variousfactory equipment and machines.

Major touch panel manufacturers areMitsubishi Electric, Keyence, and OMRON.

The main reasons for using touch panels in factories are.
1. easy to add buttons and lamps
2. monitoroperation status
3. anomalies and history of anomalies can be checked on the monitor
4. high and low temperature, vibration, shock, etc.

Chapter 2 Basic Operation of GT Designer3 (GOT2000)

Now that the installation of GT Designer3 (GOT2000) is finished, we will explain the basic operations required for screen creation.

2-1 Creating a new project.

1. Open GT Designer3 (GOT2000).

2. Select "New" from the project selection screen.

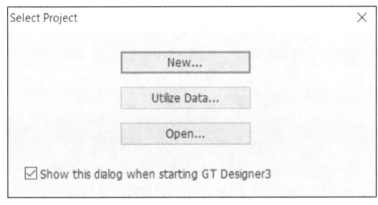

3. The New Project Wizard start screen will appear.

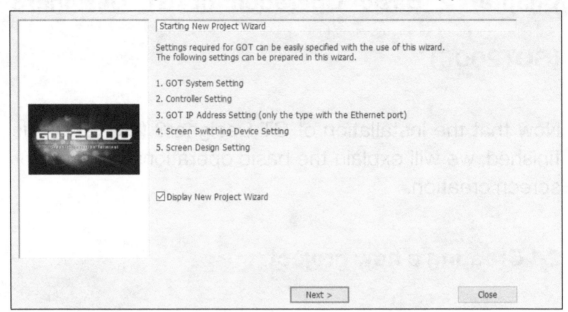

4. The GOT system setup screen will appear.

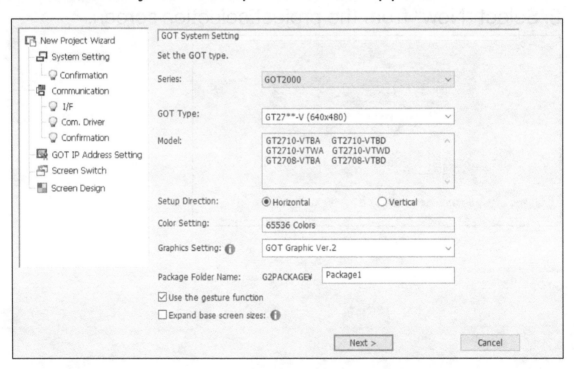

Basically, everything is fine with the default settings, but the major settings are described below.

Series : Select the series of the touch panel. In this case, GOT2000 series will be used, so no change is necessary; if GOT1000 series is used, change to "GOT1000" in the series and press "Next" to switch to the software for GOT1000 series.

GOT Type : There are several models of touch panels with different functions. Normally, the model of the touch panel to be used should be selected, but for the purpose of this screen creation, GT27**-V is used as the default setting.

Setup Direction : Set the installation orientation according to the installation condition of the touch panel. Most of them are placed horizontally, so "Horizontal" is selected this time.

Color setting :
Depending on the model, color/monochrome, etc. can be selected. In the model GT27**-V used in this study, "65536 colors" are fixed.

5. Press "Next" when the confirmation screen for GOT system settings is displayed.

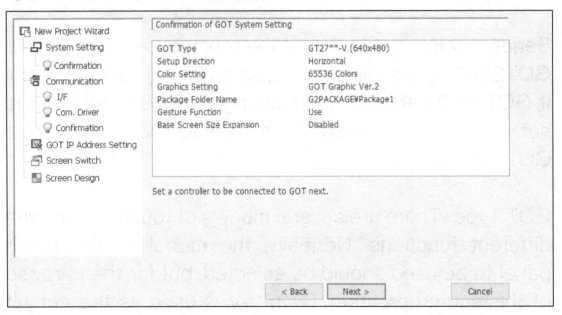

6. Select the manufacturer: "MITSUBISHI ELECTRIC" (default setting) and the controller type: "MELSEC-Q/QS" (for simulation with this model), and press "Next".

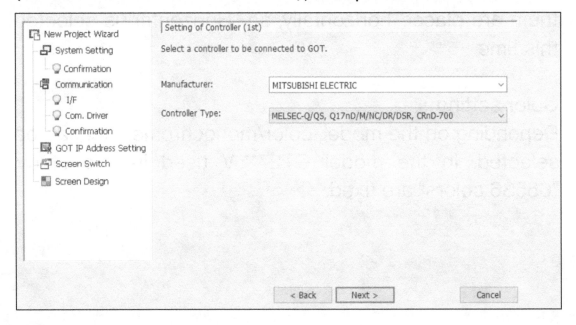

7. The same setup screen for the connected device (the first device) will be displayed.

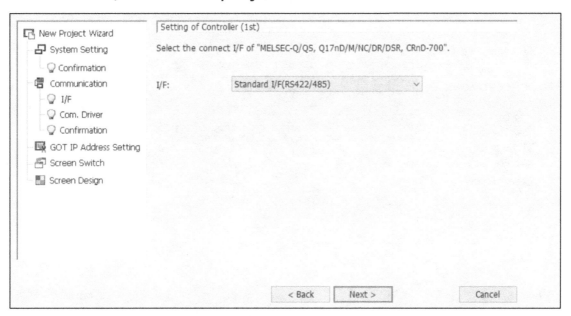

8. The same setup screen for the connected device (the first device) will be displayed, so leave "Serial" (default setting) as the communication driver and press "Next".

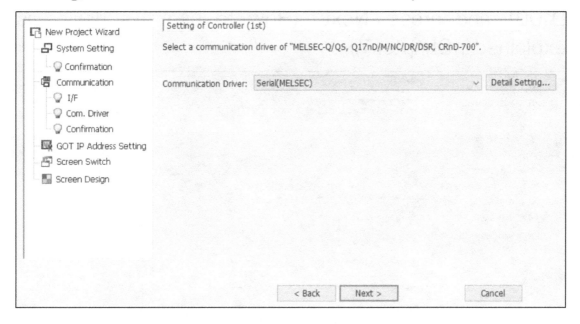

9. The confirmation screen for the connected device (first device) setting will be displayed, and press "Next".

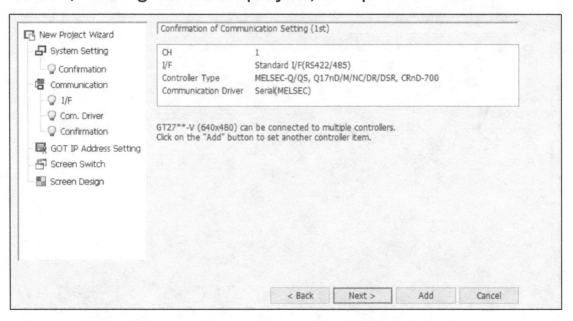

10. Next, the screen switching device setting screen will appear. By default, GD100 and GD101 are set as touch panel devices, so change GD100 to D1000 and GD101 to D1001, and press "Next". (How to use this device is explained inChapter 7.)

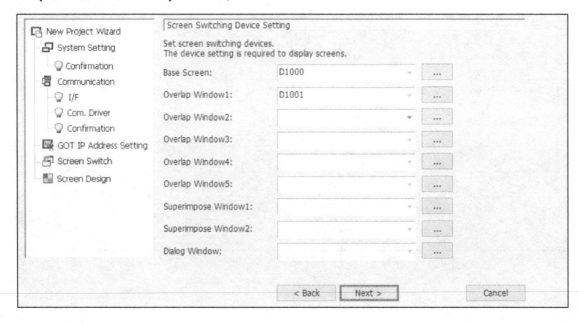

11. Next, the screen design setting screen will appear.

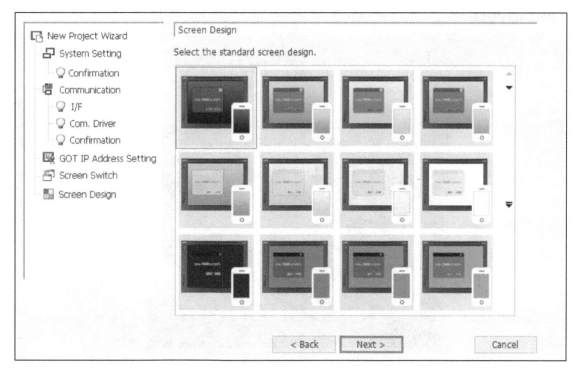

12. A confirmation screen for the system environment settings will appear.

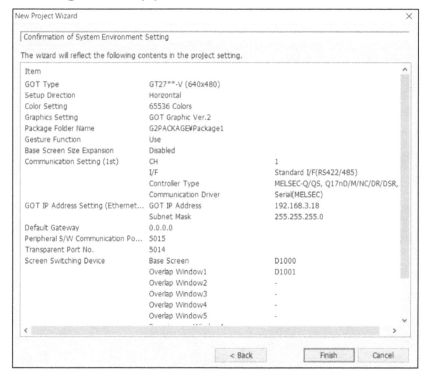

13. The following touch panel creation screen is displayed.

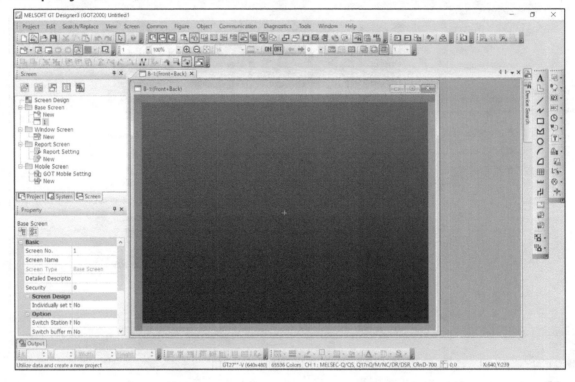

2-2 Saving Projects

1. select the "Project" tab in the upper left corner ⇒"Save As...". (If you want to overwrite the file, select "Save")

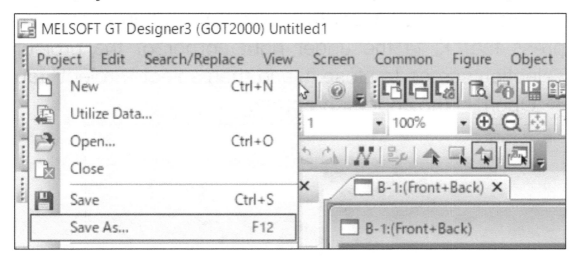

2. Name the file and press "Save"(保存).

2-3 Open a saved project.

1. Open GT Designer3 (GOT2000).

2. Select "Open..." from the project selection screen, or press the "X" in the upper right corner of the project selection screen to close it, and then select "Project" tab ⇒"Open" in the upper left corner.

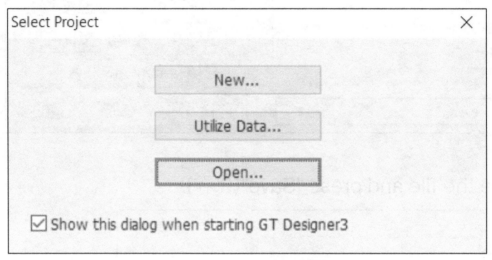

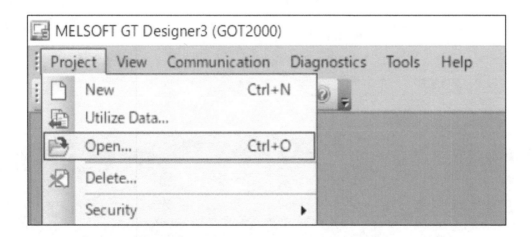

3. Select the saved file and press "Open"(開く).

You can also open a saved project by double-clicking on it.

2-4 How toCreateTouch Panel Screen

This section explains how to create atouch panel screen
using the procedure for creating a sample screen.

The screento be created is,
Lamp (Y0) lights up when switch (X0) is pressed.
Number of times lamp is lit (D0)

This sectionexplains how to use the basic components through the creation of this screen.

1. First, create a switch (X0). Select "Object" tab ⇒ Switch ⇒Bit Switch.

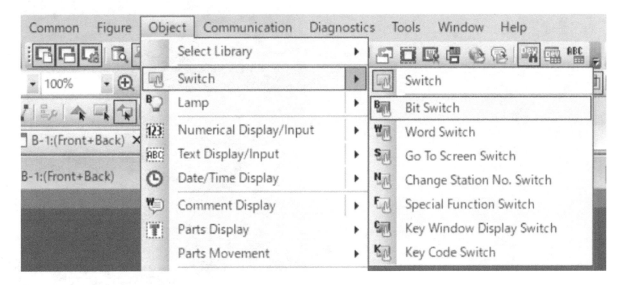

2. click on the screen to display the switch components.

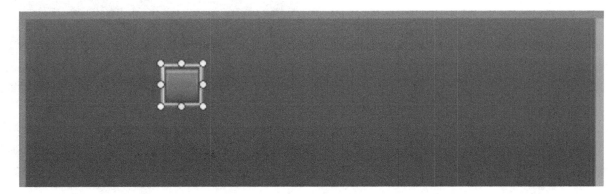

3. After clicking on the switch component, resize the bit switch by changing the width to 150 and the height to 150 in the lower left corner of thescreen.

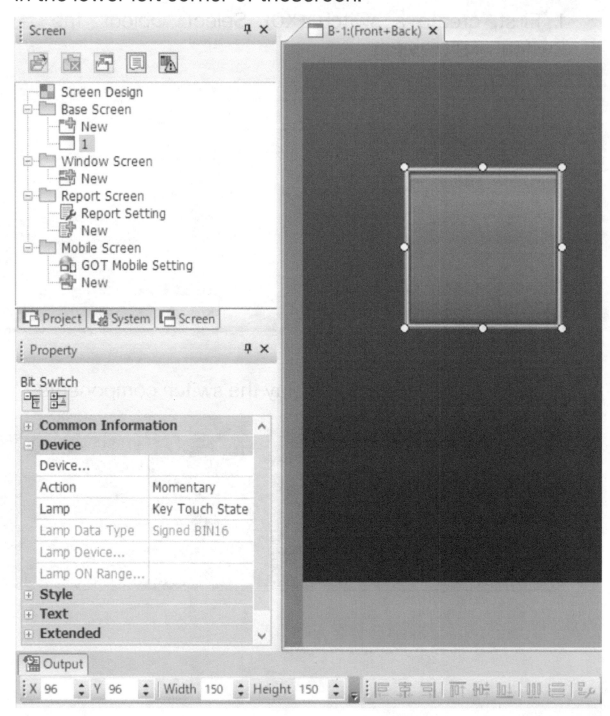

4. Configure various settings for the switch component. Double-click the switch component, and the following screen will appear.

*The device in the PLC is "X0", but on the touch panel it is displayed as "X0000". (If you put "X0" in the device, it will automatically change to X0000.)

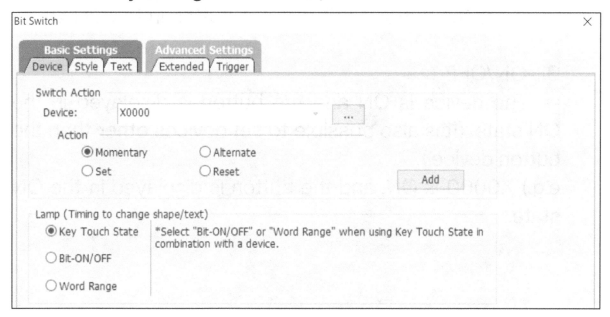

Other settings are described below. (In this case, you can leave the default settings as they are.)

[Action]

Momentary ⇒ The device is ON only while the button is pressed.

Altemate ⇒ Pressing the button while the device is OFF turning the device ON, and pressing the button while the device is ON turns the device OFF.

Set ⇒ Press the button to turn the device ON.

Reset ⇒ Press the button to leave the device OFF.

[Lamp (Timing to change shape/text)]

Key Touch State
⇒ Press the button to display the button in the ON state.

Bit-ON/OFF
⇒ The device is ON and the button is displayed in the ON state. (It is also possible to set devices other than the button device.)
e.g.) X0000 is ON, and the button is displayed in the ON state.

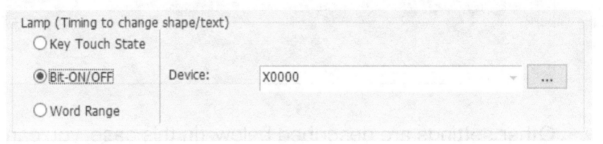

Word Range
⇒ The value of the device is displayed in the ON range and the button is displayed in the ON state.
e.g.) D10 is between 100 and 200 and the button is displayed in the ON state.

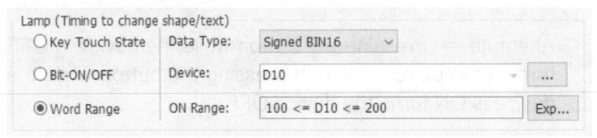

5. Next, click on the "Style" tab in the Basic Settings, and then click on "Shape..." to the right of Shape. In this screen, you can also change the color of the switches by using the shape color.

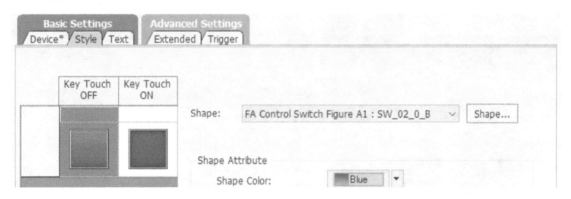

6. A list ofimages will be displayed. After selecting the key touch OFF for the displayed figure, click "1 SW_01_0_B", after selecting the key touch ON for the displayed figure, click "2 SW_01_1_B", and press "OK" in the lower right corner.

Before change

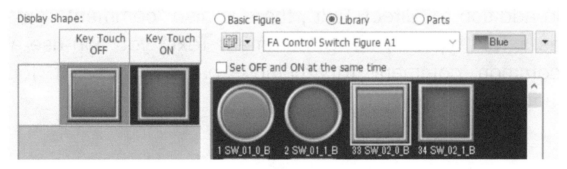

After the change

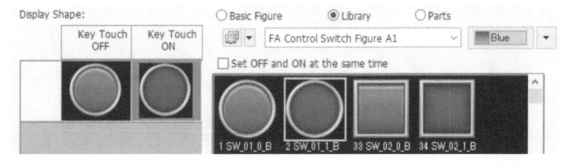

7. Next, after clicking on the "Text" tab in Basic Settings, select Key Touch OFF and uncheck "OFF=ON" in the Text Type. Then enter 22 dots for the Text size and "Switch OFF" in the text field.

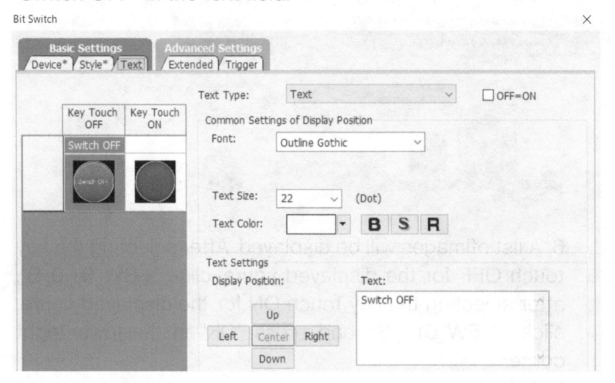

In addition to "direct Text", there is also "comment Text" as a Text type. By using "comment Text", you can use a common comment for multiple parts or use it for language switching.

8. Select "Key Touch ON," enter 22 dots for the Text size and "Switch ON" in the text field, then press "OK" in the lower right corner.

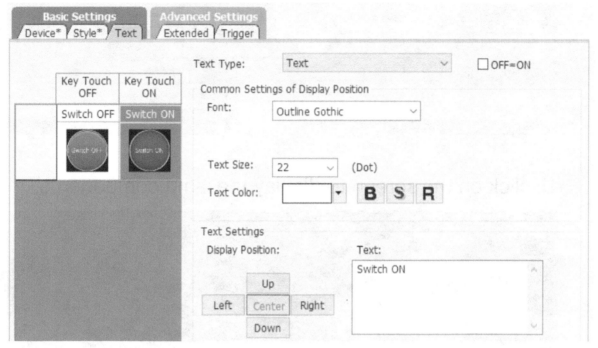

9. The switch (X0) is now complete. (Position adjustment will be done later.)

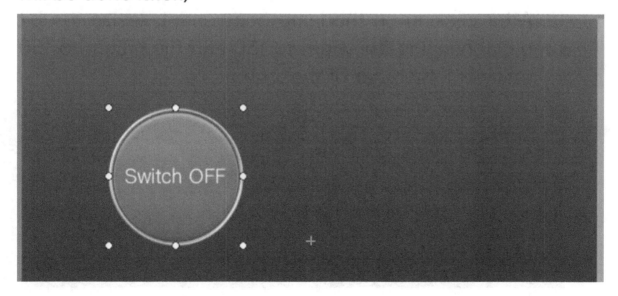

10. Next, create a Lamp (Y0). Select "Object" tab ⇒Lamp ⇒Bit Lamp.

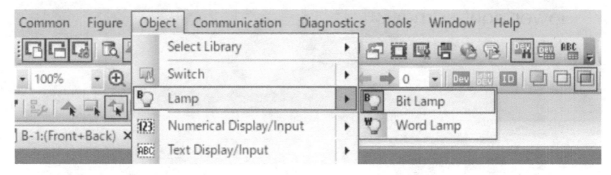

11. click on the screen to display the lamp components.

12. After clicking on the lamp component, resize the bit lamp by changing the width to 150 and the height to 150 in the lower left corner of thescreen.

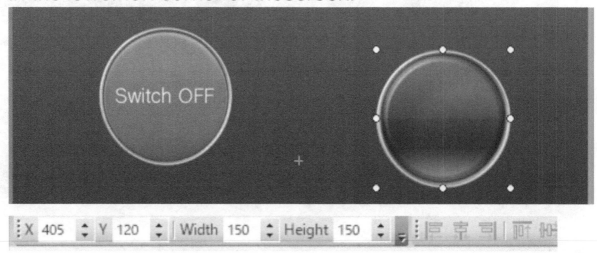

13. Set up various settings for the lamp component. Double-click the lamp component, and thefollowing screen will be displayed.

Also, after setting the Shape color to "Yellow" with the lamp OFF selected, select the lamp ON and change the shape color to "Yellow".

We will not change the shape this time, but if you want to change the shape, click on "Shape..." under Shape (P)to change it.

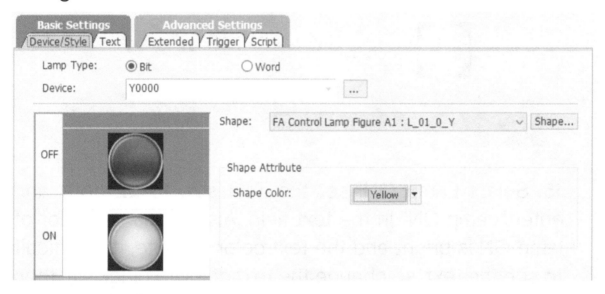

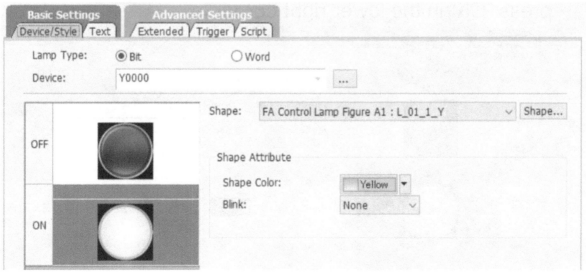

14. Next, after clicking on the "Text" tab in Basic Settings, select "Lamp OFF" and uncheck "OFF=ON" in the Text Type. Then, set the Text size to 22 dots and enter "Lamp OFF" in the text field.

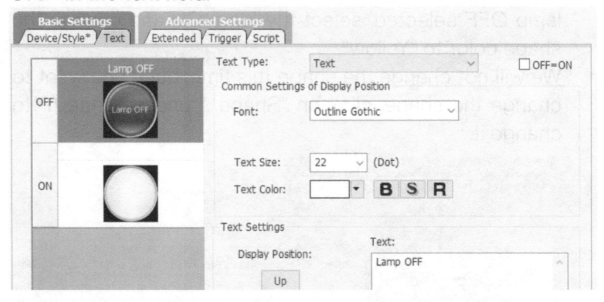

15. Select Lamp ON, set the Text size to 22 dots, and enter "Lamp ON" in the text field. Also, since the color of lamp ON is bright and the text color is white, it is difficult to see the text, so change the text color to black (0), then press "OK" in the lower right corner.

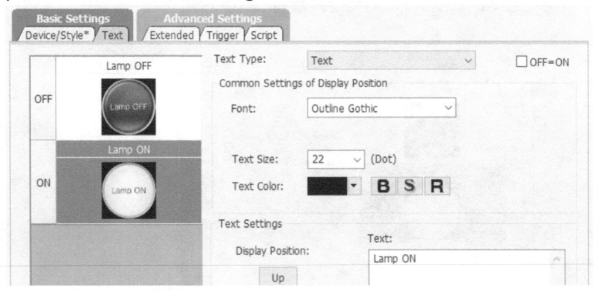

16. The lamp (Y0) is now complete. (Position adjustment will be done later.)

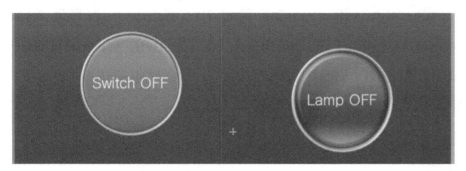

17. Next, create the lamp lighting frequency (D0). Select "Object" tab => Numerical Display/Input => Numerical Display.

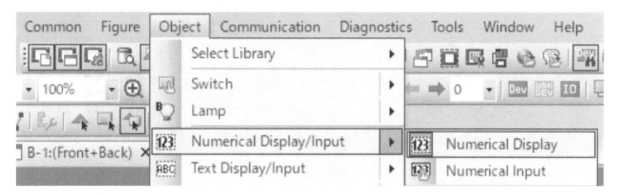

18. click on the screen to display the numeric display component.

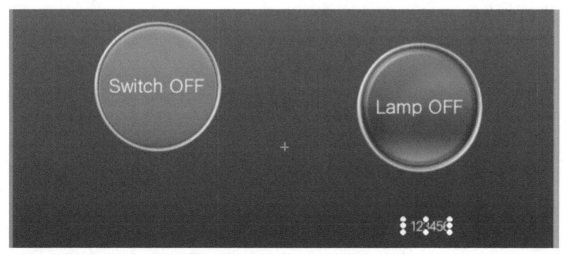

19. set the various settings for the numeric display component. Double-click the numeric display part, and the following screen will appear. Set the device to D0, the number size to 48, the alignment to center, the integer digits to 4, and the preview value to 0.

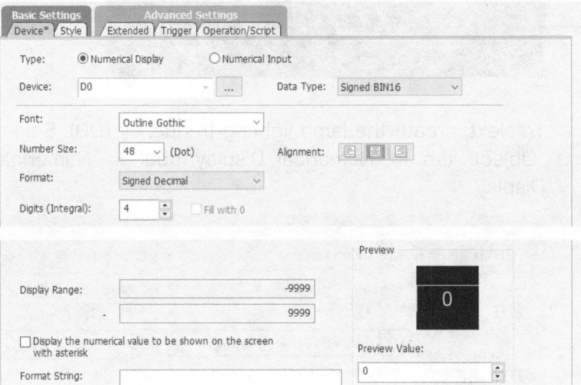

20. Next, click on the "Style" tab in the Basic Settings, then click on "Shape..." next to Shape (H). next to the shape (H).

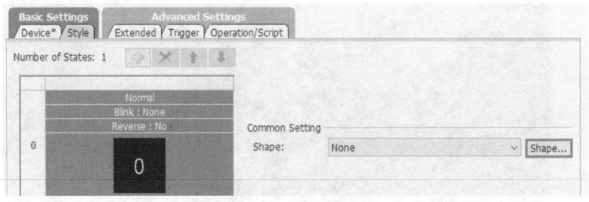

21. the image list will be displayed, select "7 Rect_7" and press "OK" at the bottom.

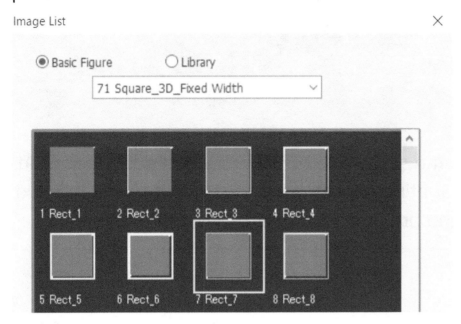

22. Then press "OK" on the bottom right. The lamp lighting frequency (D0) is now complete. (Position adjustment will be done later.)

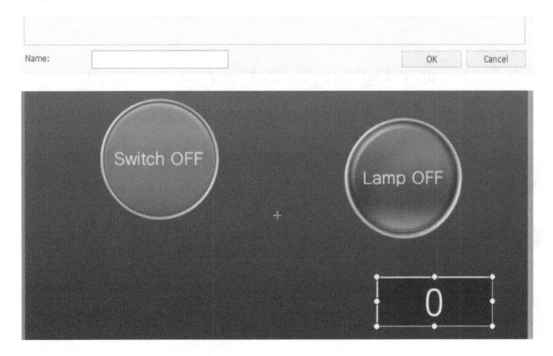

23. Finally, create the Text "Lamp Lighting Count" above the Lamp Lighting Count (D0). Select "Figure"tab => Text.

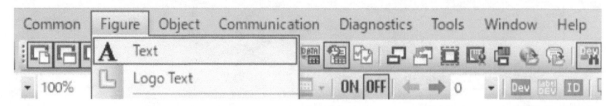

24. Click on the screen to display the text setting screen. Set "Lamp Lighting Count" as the text string and the text size to 22, and press "OK" in the lower right corner.

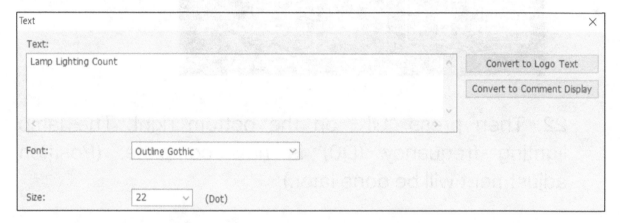

25.The Text "Lamp Lighting Count" is now complete.

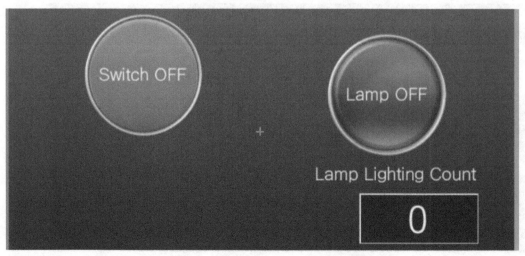

26. Finally, adjust the position of each component. Although it is possible to adjust the position of each component by moving it with theleft mouse button clicked continuously, fine adjustment is difficult, so the XY adjustment method in the lower left corner of the screen is used here.

Click on each component and set the X and Y values in the lower left corner as shown below.
Switch parts X:100 Y:100
Lamp parts X:400 Y:100
Numeric parts X:400 Y:330
Text X:400 Y:300

27. This completes the creation of the sample screen. Do not forget to save the file. (We will check the operation later in the simulation.)

Also, the following screen, which was displayed in the sample screen, has been switched ON with the ON/OFF button under the "Diagnostics" tab. You can use this function to confirm the display when it is ON/OFF. *The number of times the lamp is turned on is1 for the preview value.

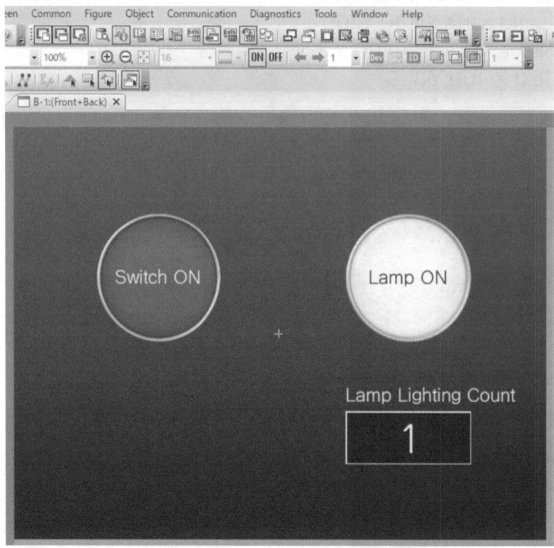

2-5 Creation Screen Title Change / NewScreen

Creation

The sample screen you created is screen number 1 with no title. To change the screen number or title, right-click on screen number 1 and select Screen Property.

This time, enter "10" for the screen number, "Switch Lamp Screen" for the title, and "Display Switch and Lamp Status and Number of Lamp Lights" for the detailed description, then press "OK" in the lower right corner.

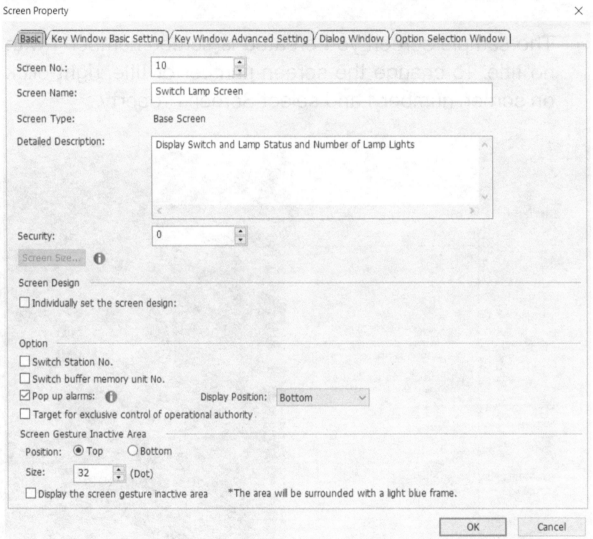

Screen number 10 and the title has been changed to "Switch Lamp Screen". These changes can also be made from the property sheet below it. (If it is hidden, you can display it by pressing Alt+1)

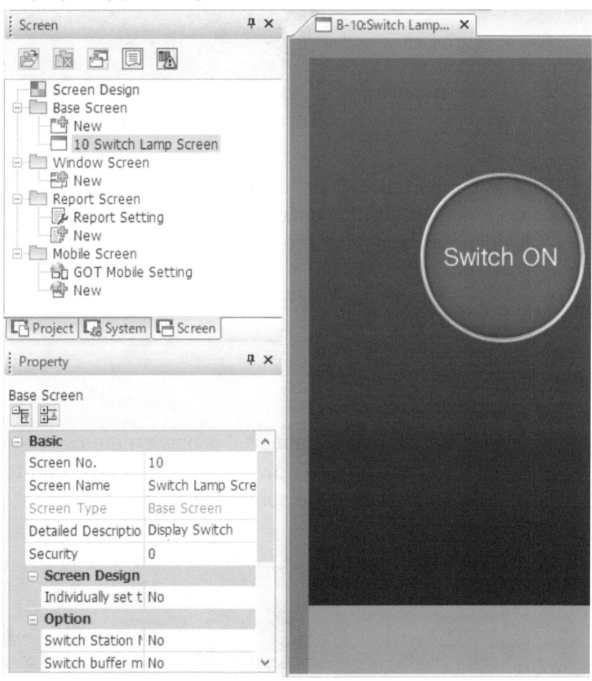

Next, we will explain how to create a new screen. First, double-click "New" under the Base Screen folder of the screen.

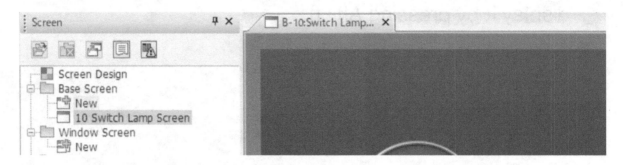

Screen Properties will be displayed." Enter "11" for the screen number and "Question 1" for the title, then press "OK" in the lower right corner.

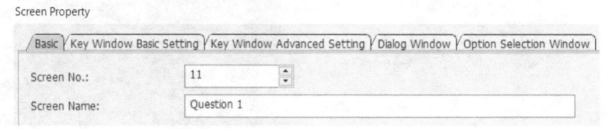

A new Screen 11, Title "Question 1" is now created.
The sample screen is now complete. Let's create Exercises 1 and 2 using the same procedure.

Exercise 1 (Shutter display)

When the shutter open button (X1) or shutter close button (X2) is pressed, the shutter status lamp (Y1) changes from "Shutter is stopped" to "Shutter is in operation". Create this screen for Screen No. 11, Question 1. (The operation will be checked in the simulation in Chapter 4.)
*The PLC program to turn on Y1 when X1 or X2 is turned on is explained in Chapter 4.

[OFF state].　　　　　　　　　[ON state].

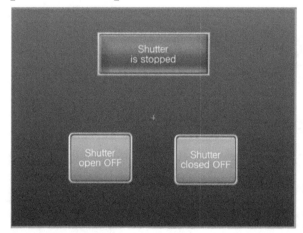 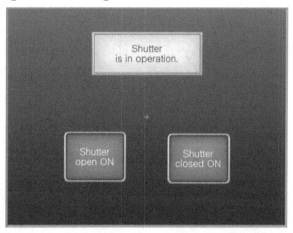

Use the following settings for each component.

Shutter open button (X1)
Object: Bitswitch   Device: X0001
OFF graphic color: green
OFF shape: FA control switch
Figure A1 71 SW_03_0_G
ON shape color: green   ON shape: FA control switch
Figure A1 72 SW_03_1_G
Text type OFF=ON check: OFF   Text size: 22
The Text OFF: Shutter open OFF.
ON Text: Shutter open ON.
Part location and size
(Displayed in the lower left corner after clicking on a part)
X:130 Y:270 W:145 H:110

Shutter close button (X2)
Object: Bitswitch   Device: X0002
OFF graphic color: green
OFF shape: FA control switch
Figure A1 71 SW_03_0_G
ON shape color: green   ON shape: FA control switch
Figure A1 72 SW_03_1_G
Text type OFF=ON check: OFF   Text size: 22
The Text OFF: Shutter closed OFF.
ON Text: Shutter closed ON.
Part location and size
(Displayed in the lower left corner after clicking on a part)
X:370 Y:270 W:145 H:110

Shutter status lamp (Y1)

Object: Bit lamp   Device: Y0001

OFF shape color: water-based

OFF shape: FA control lamp

Figure A1 75 L_03_1_LB

ON shape color: water-based

ON Shape: FA Control Lamp

Figure A1 76 L_03_0_LB

Text type OFF=ON check: OFF   Text size: 22

Text OFF: Shutter is stopped.

Text color: white

Text ON: Shutter is in operation.

Text color: black

Part location and size

(Displayed in the lower left corner after clicking on a part)

X:190 Y:50 W:260 H:100

Exercise 1 Solution

Shutter open button (X1)　Object: Bitswitch

Basic Settings (Devices) Device: X0001

Basic Settings (Style)
OFF shape color: Green
OFF shape: FA Control Switch Figure A1 71 SW_03_0_G

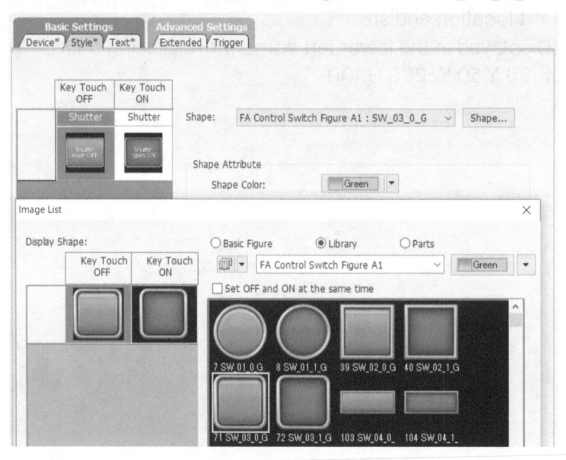

ON shape color: Green
ON shape: FA Control Switch Figure A1 72 SW_03_1_G

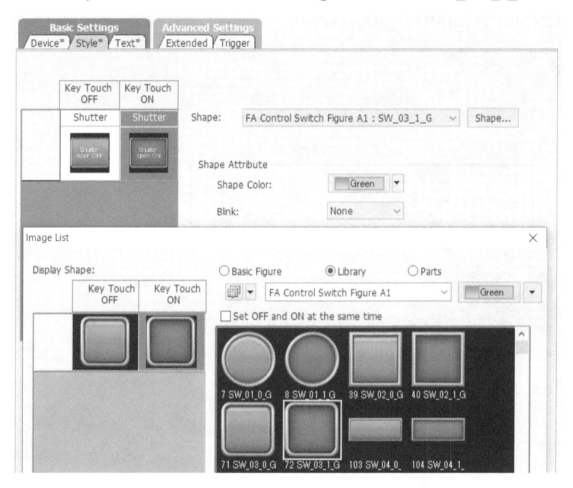

[Basic settings (text)]
Text type OFF=ON check: OFF   Text size: 22
Text OFF: Shutter open OFF.

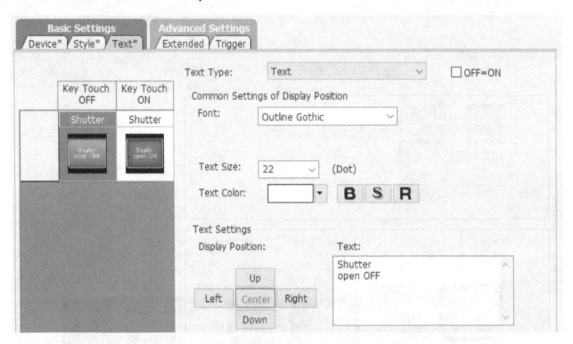

Text ON: Shutter open ON.

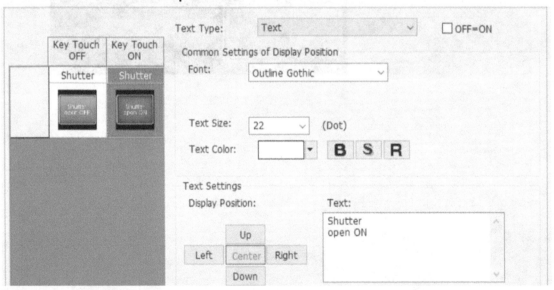

Part location and size
(Displayed in the lower left corner after clicking on a part)
X:130 Y:270 W:145 H:110

Shutter close button (X2)

The same procedure can be used for the shutter open button (X1), but sincethe settings for the open and close buttons are similar, we will use the Copy ⇒Paste method here.

Right click on the shutter open button (X1) => select copy.

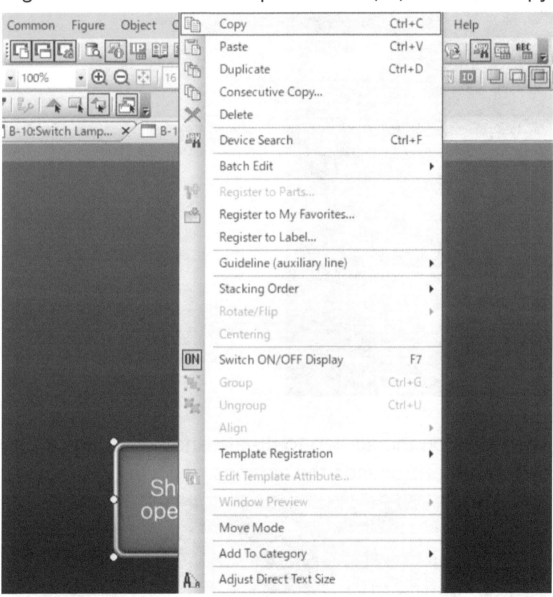

Right-click on the screen ⇒ select Paste.

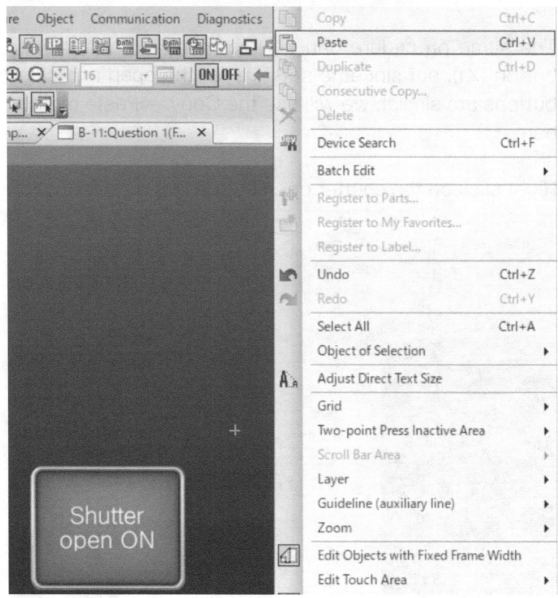

Then click on the screen to create another shutter open button (X1) and change the settings.

Basic Settings (Devices)
Device: Changed to X0002

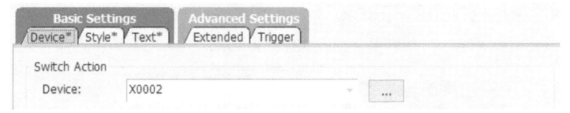

[Basic settings (text)]
The Text OFF: Shutter closed OFF.

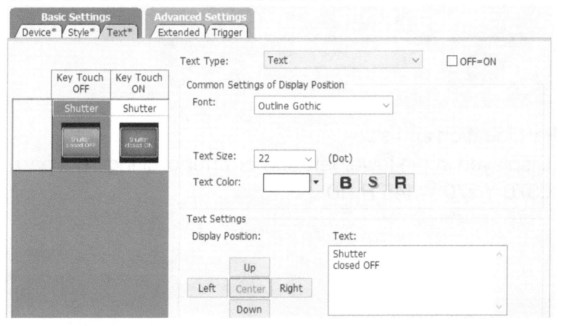

Text ON: Shutter closed ON.

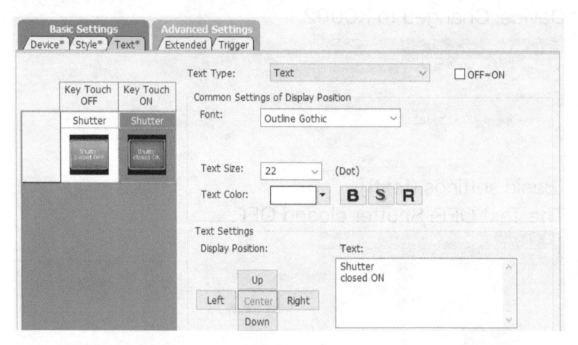

Part location and size
(Displayed in the lower left corner after clicking on a part)
X:370 Y:270 W:145 H:110

Shutter status lamp (Y1)
Object: Bit Lamp
Basic Settings (Device/Style)
Device: Y0001
OFF shape color: Cyan
OFF shape: FA Control Lamp Figure A1 75 L_03_1_LB

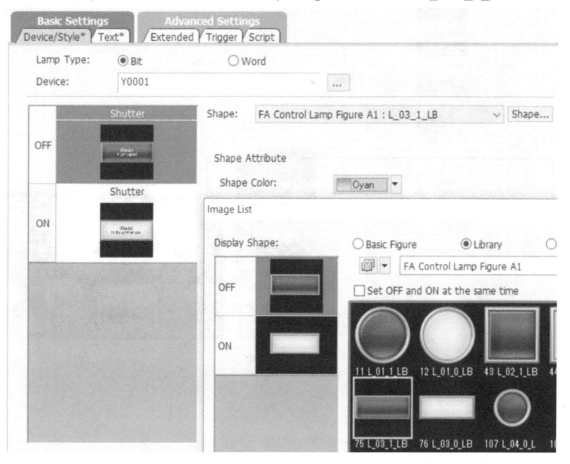

ON shape color: Cyan
ON Shape: FA Control Lamp Figure A1 76 L_03_0_LB

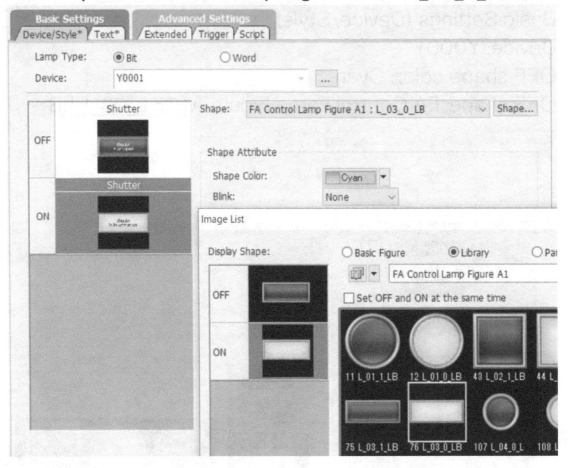

[Basic settings (text)]
Text type OFF=ON check: OFF
Text size: 22
Text OFF: Shutter is stopped.
Text color: white

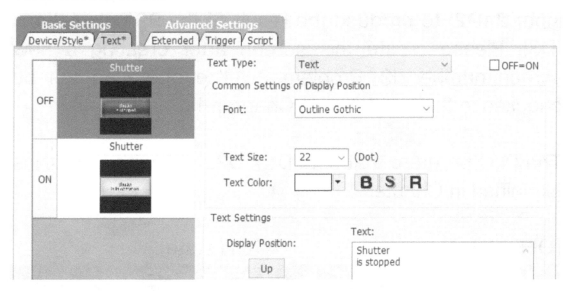

Text ON: Shutter is in operation.
Text color: black

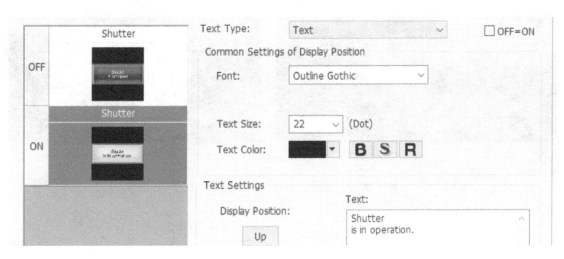

Part location and size
(Displayed in the lower left corner after clicking on a part)
X:190 Y:50 W:260 H:100

Exercise 2 (Calculation)

Press the Perform Calculation button (X3) to multiply the single-digit number input 1 (D1) by the single-digit number input 2 (D2) to produce the result of the 99 calculations (D3). Please create this screen after creating a new screen number 12, Problem 2. (The operation will be checked in the simulation in Chapter 4.)

The PLC program that sets D1 x D2 = D3 with *X3 ON is explained in Chapter 4

[OFF state].   [ON state].

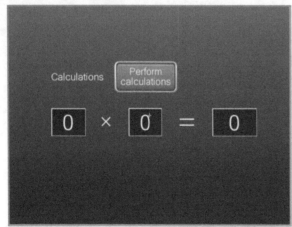   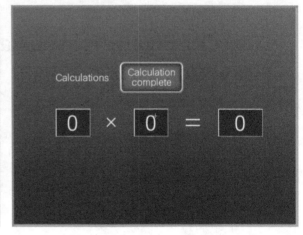

Use the following settings for each component.

Calculations (Texts)
Figure ⇒Text Text size: 22
Figure position and size
(Displayed in the lower left corner after clicking on a part)
X:100 Y:140 Width:110

... x (text)
Figure ⇒Text Text size: 48
Figure position and size
(Displayed in the lower left corner after clicking on a part)
X:210 Y:225 Width:26

... = (Text)
Figure ⇒Text Text size: 48
Figure position and size
(Displayed in the lower left corner after clicking on a part)
X:380 Y:230 Width:48

Calculation button (X3)
Object: Bit switch
Device: X0003
OFF shape color: reddish
OFF shape: FA Control Switch Figure A1 131 SW_05_0_R
ON shape color: reddish
ON shape: FA Control Switch Figure A1 132 SW_05_1_R
Text type OFF=ON check: OFF
Text size: 22
Text OFF: Perform calculations.
Text ON: Calculation complete.
Part location and size
(Displayed in the lower left corner after clicking on a part)
X:240 Y:130 W:145 H:50

One-digit numeric input 1 (D1)
Object: Numerical input
Device: D1 Text size: 48
Alignment: Centered   Digits: 1
Preview value:0
Shape:71 Square_Solid_Border_Width_Fixed_7 Rect_7
Part location and size
(Displayed in the lower left corner after clicking on a part)
X:100 Y:225 W:78 H:62

One-digit numeric input 2 (D2)

Object: Numerical input

Device: D2 Text size: 48

Alignment: Centered   Digits: 1

Preview value:0

Shape:71 Square_Solid_Border_Width_Fixed_7 Rect_7

Part location and size

(Displayed in the lower left corner after clicking on a part)

X:270 Y:225 W:78 H:62

Result of calculations (D3)

Object: Numerical Display

Device: D3 Text size: 48

Alignment: Centered   Digits: 2

Preview value:0

Shape:71 Square_Solid_Border_Width_Fixed_7 Rect_7

Part location and size

(Displayed in the lower left corner after clicking on a part)

X:460 Y:225 W:102 H:62

Exercise 2 Solution

After double-clicking New on the base screen, enter screen number: 12, title: Problem 2, and press "OK" to create a new screen.

Calculations (Texts)
Figure ⇒ Text Text size: 22

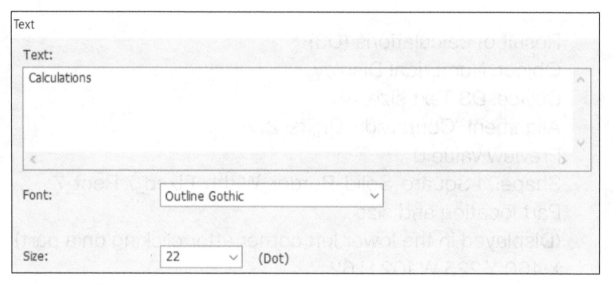

Figure position and size
(Displayed in the lower left corner after clicking on a part)
X:100 Y:140 Width:110

... x, = (Text)

Figure ⇒ Text Text size: 48

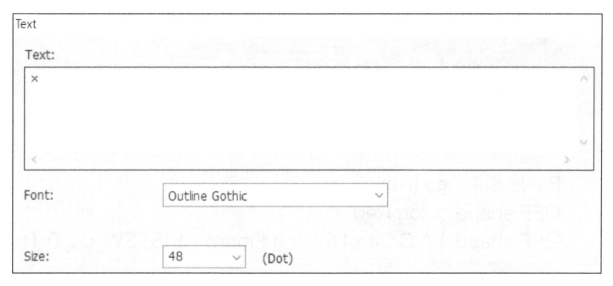

Figure position and size

(Displayed in the lower left corner after clicking on a part)

X:210 Y:225 Width:26

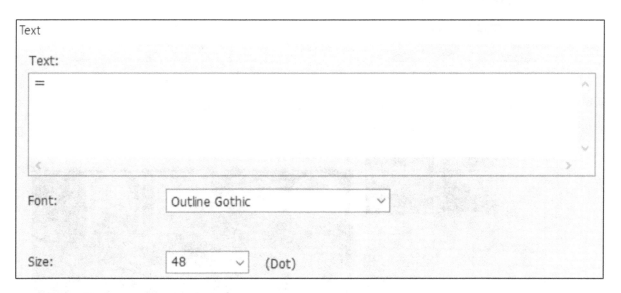

Figure position and size

(Displayed in the lower left corner after clicking on a part)

X:380 Y:230 Width:48

Calculation button (X3)
Object: Bit switch
Basic Settings (Devices)　Device: X0003

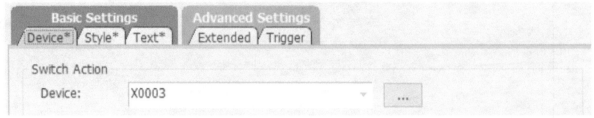

Basic Settings (Style)
OFF shape color: Red
OFF shape: FA Control Switch Figure A1 131 SW_05_0_R

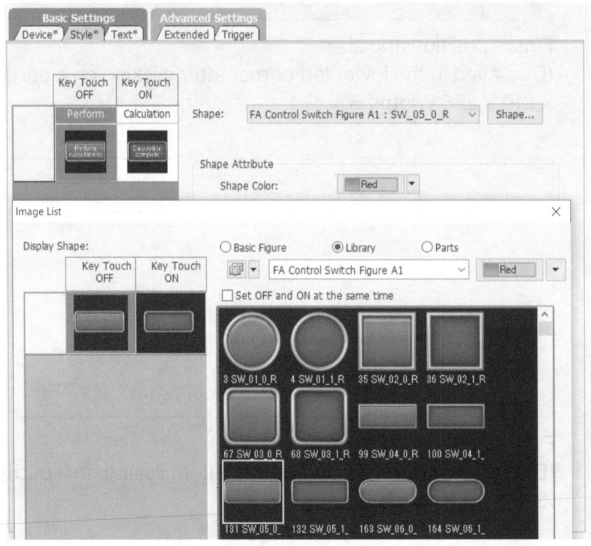

ON shape color: Red
ON shape: FA Control Switch Figure A1 132 SW_05_1_R

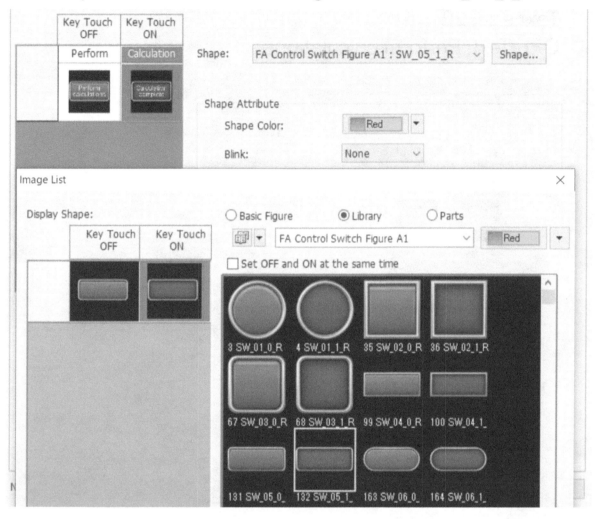

[Basic settings (Text)]
Text type OFF=ON check: OFF
Text size: 22
OFF Text: Perform calculations.

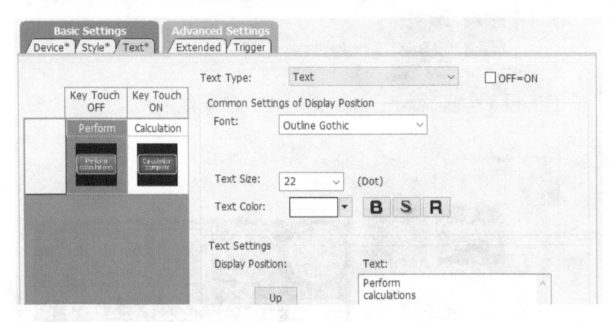

ON Text: Calculation complete

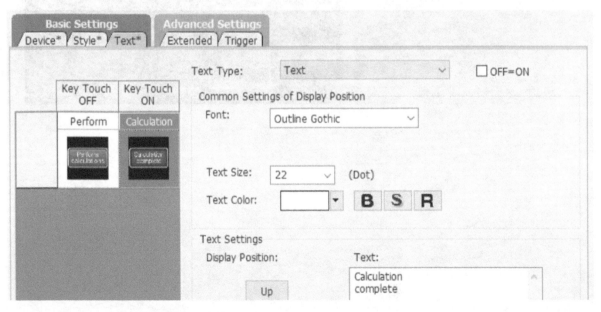

Component location and size
(Displayed in the lower left corner after clicking on a part)
X:240 Y:130 W:145 H:50

One-digit numeric input 1 (D1)
Object: Numerical input
Basic Settings (Device)
Device: D1   Number size: 48
Alignment: Centered   Digits: 1
Preview value:0

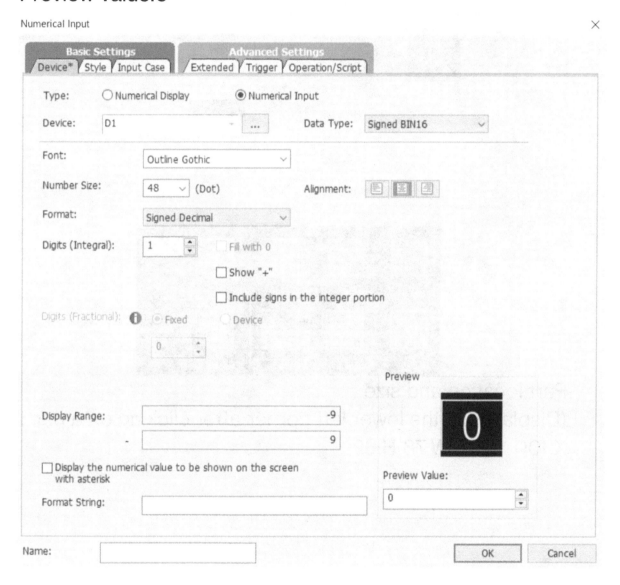

Basic Settings (Style)
Shape:71 Square_Solid_Border_Width_Fixed 7 Rect_7

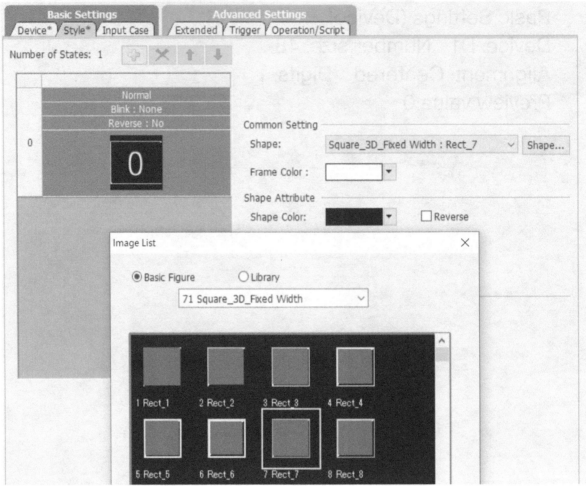

Part location and size
(Displayed in the lower left corner after clicking on a part)
X:100 Y:225 W:78 H:62

One-digit numeric input 2 (D2)
⇒Right-click on a part of one-digit numerical input 1 ⇒
Right-clickafter copying ⇒Right-click ⇒Paste
Basic Settings (Device)　Device: Changed to D2

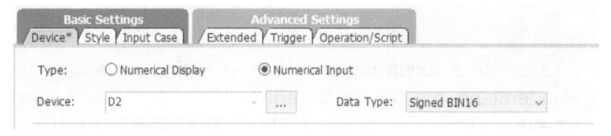

Part location and size
(Displayed in the lower left corner after clicking on a part)
X:270 Y:225 W:78 H:62

Result of calculation (D3)
⇒Right-click on a part of one-digit numerical input 1 =>
Copy, then right-click => Paste
Basic Settings (Device)
Type: Numerical Display　Device: D3
Digits: Changed to 2

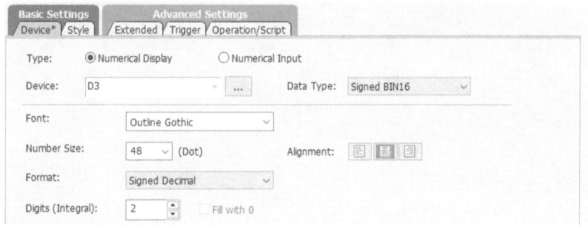

Part location and size
(Displayed in the lower left corner after clicking on a part)
X:460 Y:225 W:102 H:62

Chapter3 About Devices

3-1 Bit and word

To create a touch panel screen, it is first necessary to understand the "devices" required for signal exchange between the touch panel and PLC. The "device" can be a "bit device" or a "word device". First, "bit" in "bit device" and "word" in "word device" are explained. A "bit" is the smallest unit that a computer can handle and canrepresent 0 or 1. A "word" can handle large numbers such as 1000 or 10000 by using multiple bits (16 bits).

3-2 What is a bit device?

Bit devices include X (input relay), Y (output relay), M (auxiliary relay), etc. Since they are "bits," they are represented by 0 or 1. For example, a device can be represented as OFF (0) when the switch is not pressed, ON (1) when the switch is pressed, and so on.

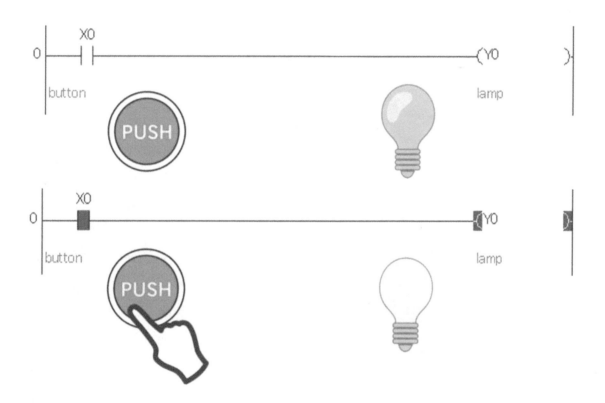

3-3 What is a word device?

Word devices include D (data register) and W (link register), which are "word (word)" and therefore represent large numbers such as 1000 and 10000.
For example, if you put 500 yen in a vending machine, the amount displayed will read 500. This device can store such numerical values.

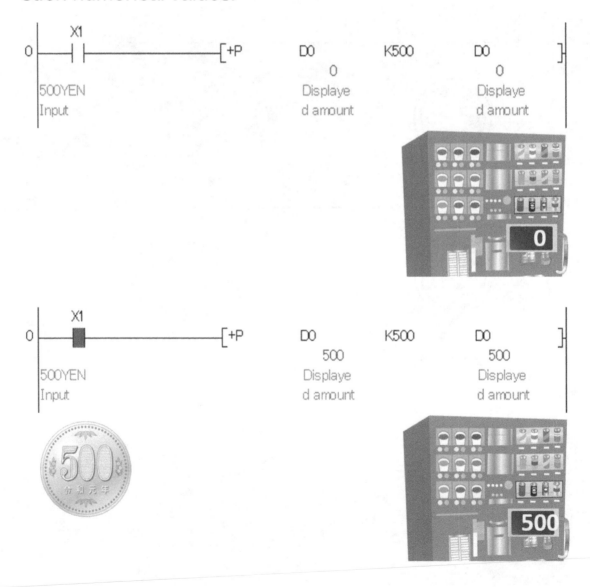

3-4 X (input relay), Y (output relay)

Next, each device is described. X (input relay) is a device that receives input from the outside (switch or sensor) and Y (output relay) is a device that outputs to the outside (lamp or motor). Also, a number is added after the device to identify what the signal is, such as X0, X1, and X2. For example, in touch-type automatic doors, the door opens when touched by hand. Withno touch, X2 is off, so Y2 is also off.

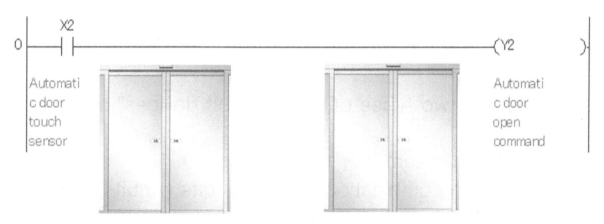

When touched, X2 is turned on and Y2, which is connected to it, is also turned on and the door opens.

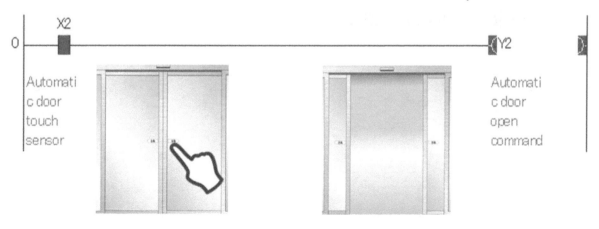

Thus, a device representing an input signal from the outside is represented by X (input relay) and a device representing an output signal to the outside by Y (output relay).

3-1 to 3-4 Summary

There are two types of devices: "bit devices" and "word devices".

Bits can represent 0 or 1.

A word is a combination of several bits (16 bits) that can represent a number from -32768 to 32767.

X (input relay) is adevice that receives inputs from the outside (switches and sensors).

Y (output relay) is adevice that provides output to the outside (lamp or motor).

3-5 Touch Panel Switch Types

In the description of the device, I mentioned that the lamp lights up when the switch is pressed, but there are four types of touch panel switches.
As explained in the section on how to create screens, the four types of switches are explained againbecausethis knowledge is necessary to understand the subsequent explanations.

Momentary
⇒Switch that stays ON only while the button is pressed.

Altemate
⇒Switch that turns ON when pressed in OFF state and turns OFF when pressed in ON state

Set
⇒Switch that remains ON when button is pressed

Reset
⇒Switch that remains OFF when button is pressed

Let's start with an example of something familiar.
For example, the buttons of crane games often seen in game arcades are "Momentary" becausethey move only as long as they are pressed.

Also, anything thatturns the monitor on when pressed once and off when pressed again, like the power button on a smartphone, is "Altemate".

The "low"(弱), "medium"(中), and "high"(強) buttons on the fan are "Set" because they remain pressed and hold the ON state when pressed.

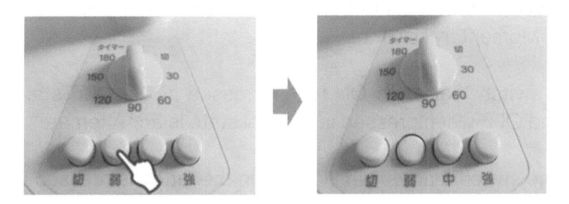

The "off" button on the fan is a "Reset" because the "low"(弱), "medium"(中), and "high"(強) buttons are reset and return to the off state.

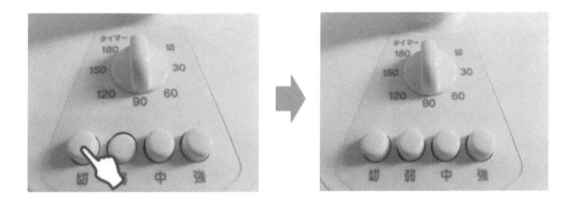

In explaining touch panel screen creation in the future, many switches will appear, but the most frequently used switch operation settings are "Momentary" and "Altemate. In the descriptions that follow, we will specify which operational settings are used, so please keep in mind the characteristics of each switch.

3-5 Summary

A switch that turns ON only while the button is pressed is a "Momentary" switch.

A switch that turns ON when pressed in the OFF state and OFF when pressed in the ON state is an "Altemate".

The switch that remains ON when thebutton is pressed is "Set".

The switch that remains OFF when thebutton is pressed is "Reset".

Exercise 3 (Type of switch)

Switch 1 (X10) turns ON lamp 1 (Y10) only while it is pressed. The PLC program to turn ON Y10 when X10 is ON is described in Chapter 4.

Switch 2 (X11) turns lamp 2 (Y11) ON when pressed once and turns lamp 2 (Y11) OFF when pressed again. The PLC program to turn ON Y11 when X11 is ON is described in Chapter 4.

Create this screen using the parts from Screen 10 (Switch Lamp Screen) after creating a new Screen #13, Question 3. (Confirm the operation by simulation in Chapter 4.)

[OFF state].   [ON state].

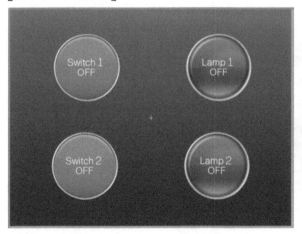
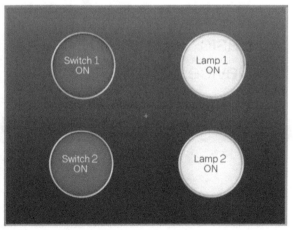

Any style, size, and position are acceptable.

Exercise 3 Solution

After double-clicking New on the base screen, enter screen number: 13, title: Problem 3, and press "OK" to create a new screen.

Switch 1 (X10)
Select the switch on screen 10, right click => select copy, then right click => paste on screen 13, click on screen

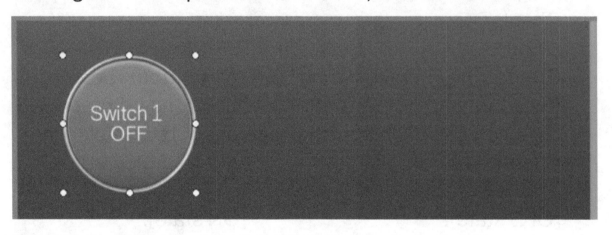

Basic Settings (Device)
Device: X0010
Action: Momentary

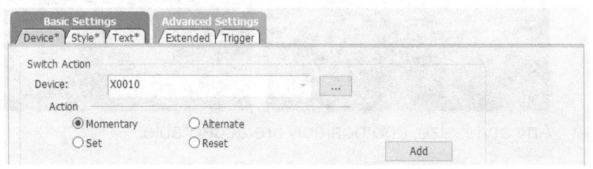

[Basic settings (text)]
Text OFF: Switch 1 OFF

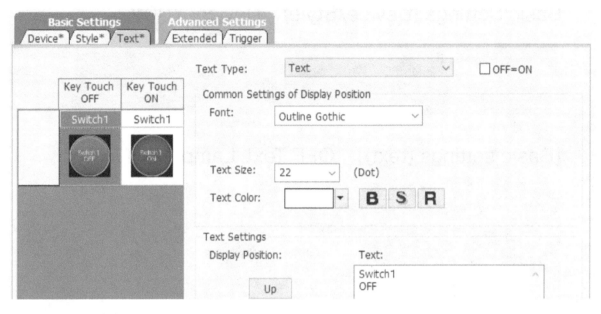

Text ON: Switch 1 ON

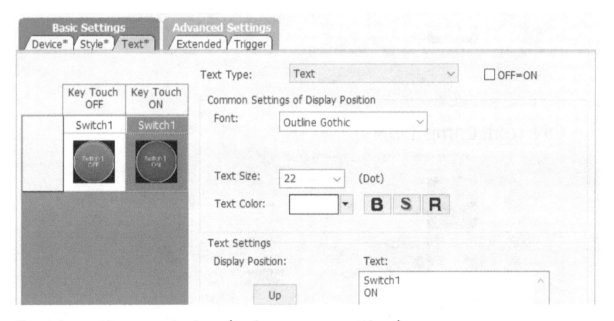

Part location and size (reference setting)
(Displayed in the lower left corner after clicking on a part)
X:100 Y:50 Width:150 Height:150

Lamp 1 (Y10)
Copy => Paste the lamp on screen 10
Basic Settings (Device/Style)　Device: Y0010

[Basic settings (text)]　OFF Text: Lamp 1 OFF

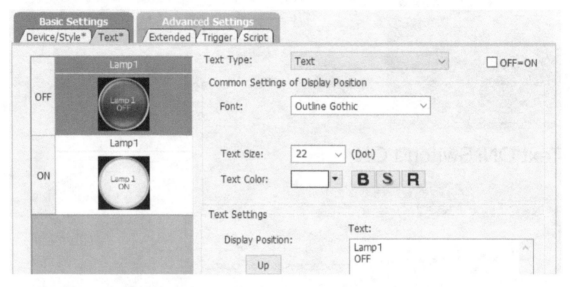

ON Text: Lamp 1 ON

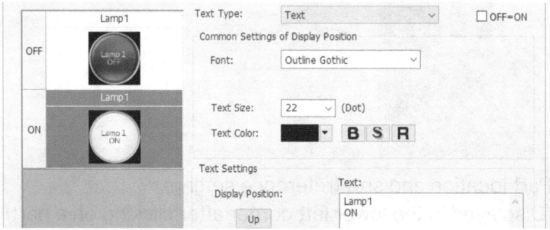

Part location and size (reference setting)
(Displayed in the lower left corner after clicking on a part)
X:400 Y:50 Width:150 Height:150

Switch 2 (X11)
Copy => Paste Switch 1
Basic Settings (Devices)
Device: X0011   Action: Altemate
Lamp function: Bit-ON/OFF   Lamp device: X0011

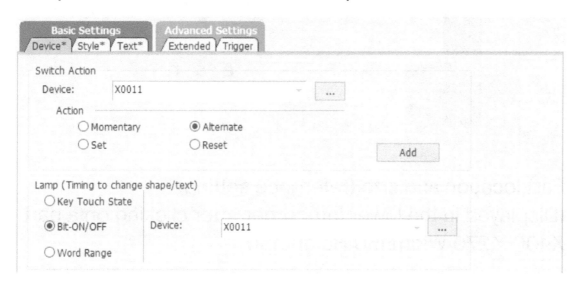

[Basic settings (text)]
Text OFF: Switch 2 OFF

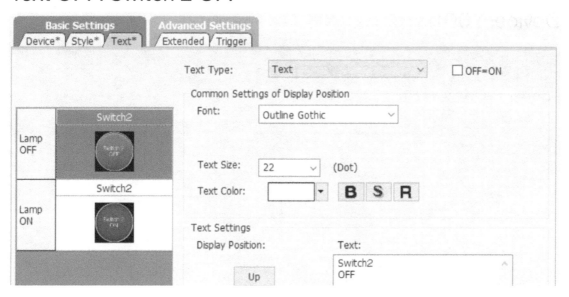

Text ON: Switch 2 ON

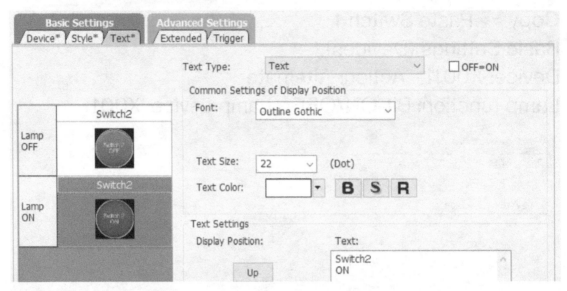

Part location and size (reference setting)
(Displayed in the lower left corner after clicking on a part)
X:100 Y:270 Width:150 Height:150

Lamp 2 (Y11)
Copy => Paste Lamp 1
Basic Settings (Device/Style)
Device: Y0011

[Basic settings (text)]
OFF Text: Lamp 2OFF

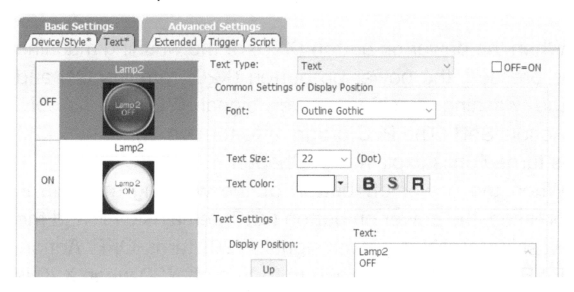

ON Text: Lamp 2 ON

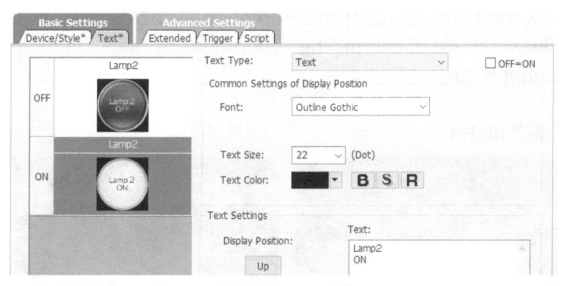

Part location and size (reference setting)
(Displayed in the lower left corner after clicking on a part)
X:400 Y:270 Width:150 Height:150

Exercise 4 (Washing machine power supply)

When the power on button (X20) of the washing machine is pressed, the power on button (X20) remains ON and the washing machine power signal (Y20) turns ON. Action: Set] *The PLC program to turn on Y20 when X20 is turned on is explained in Chapter 4.

When the power off button of the washing machine is pressed, the power on button (X20) remains OFF and the washing machine power signal (Y20) turns OFF. Action: Bit Reset] *The PLC program to turn off Y20 when X20 is turned off is explained in Chapter 4.

Create this screen using the parts from Screen 10 (Switch Lamp Screen) after creating a new Screen #14, Question 4. (Confirm the operation by simulation in Chapter 4.)

[OFF state].   [ON state].

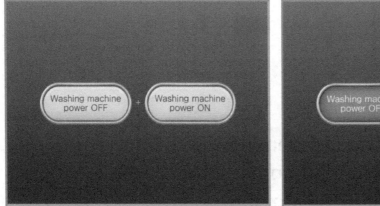

Any style, size, and position are acceptable.

Exercise 4 Solution

After double-clicking New on the base screen, enter screen number: 14, title: Problem 4, and press "OK" to create a new screen.

Power on button of washing machine (X20)

Copy ⇒Paste the switches on screen 10

Basic Settings (Devices)
Device: X0020
Action: Set
Lamp function: Bit ON/OFF
Lamp function device: Y0020

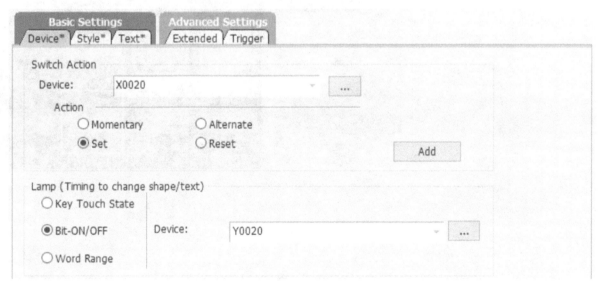

Basic Settings (Style)] (Reference setting)
OFF shape color: white
OFF shape: FA Control Switch Figure A1 185 SW_06_0_W

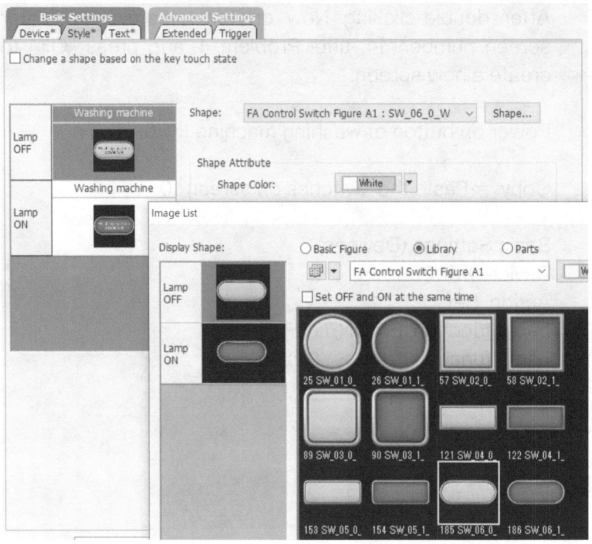

ON shape color: white
ON shape: FA Control Switch Figure A1 186 SW_06_1_W

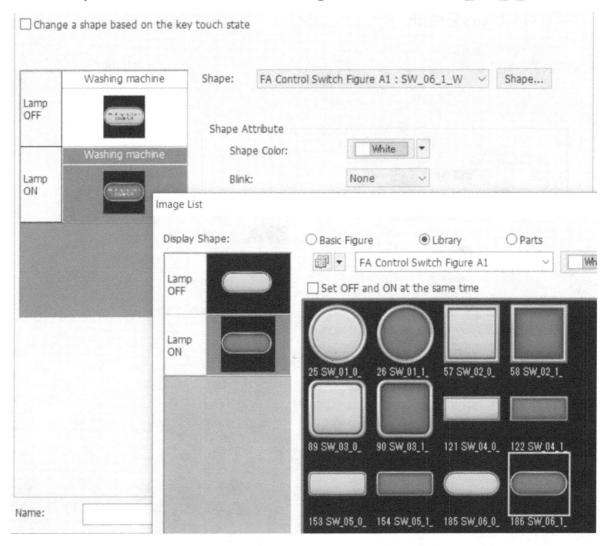

[Basic settings (text)]
OFF text: Washing machine power ON
Text color: Black.

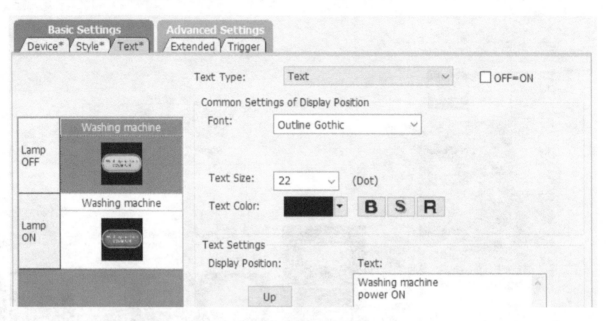

ON text: Washing machine power ON
Text color: White.

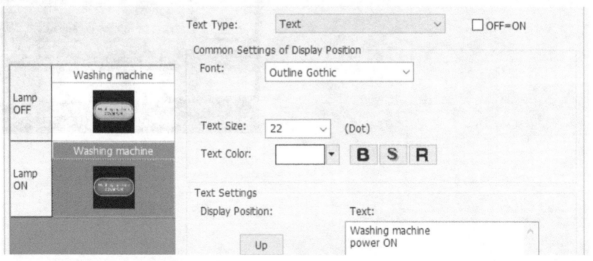

Part location and size (reference setting)
(Displayed in the lower left corner after clicking on a part)
X:340 Y:190 W:225 H:100

Power off button on washing machine (X20)

Copy ⇒ Paste the power on button of the washing machine

Basic Settings (Devices)
Device: X0020
Action: Reset.
Lamp function: key touch status

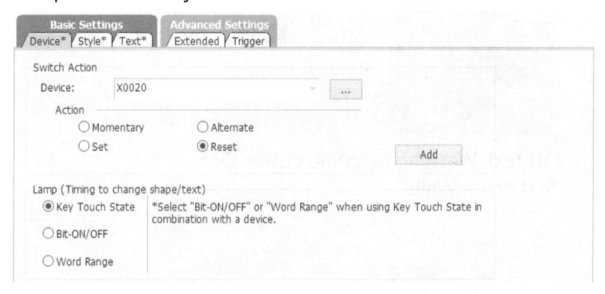

[Basic settings (text)]
OFF text: Washing machine power OFF.
Text color: Black.

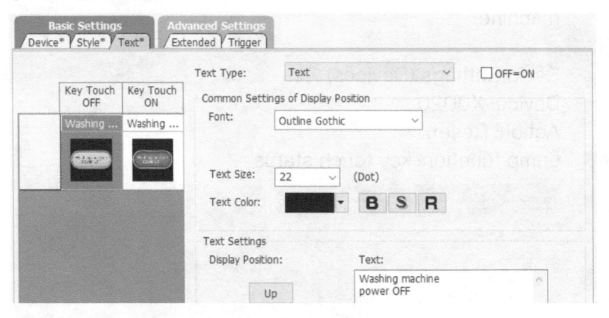

ON text: Washing machine power OFF.
Text color: White.

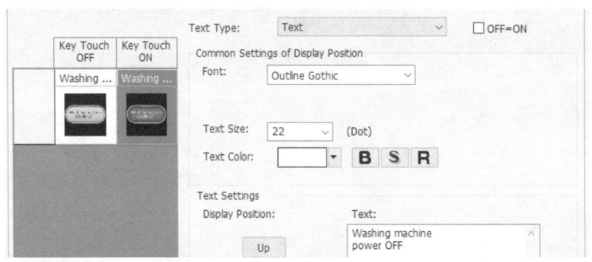

Part location and size (reference setting)
(Displayed in the lower left corner after clicking on a part)
X:60 Y:190 W:225 H:100

Chapter 4: Checking the simulation operation of the creation screen.

So far, we have created a switch lamp screen and four practice questions, and you can use the simulation function to check if those screens work correctly.

Screen creation software GT Designer3 (GOT2000) and PLC programming software GX Works2 are used to check the operation in the simulation. By using these two software programs, it is possible to perform realistic operation checks as if the touch panel were being operated.

After this, simulate the switch lamp screen and the four practice questions.

4-1 Creating a new project.

1. Open GX Works2.

2. Select the "Project" tab in the upper left corner ⇒ "New".

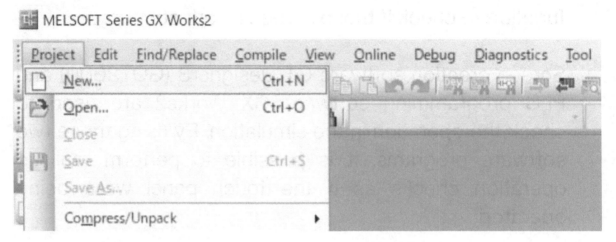

3. A screen for selecting the series, model, project type, and program language will appear.

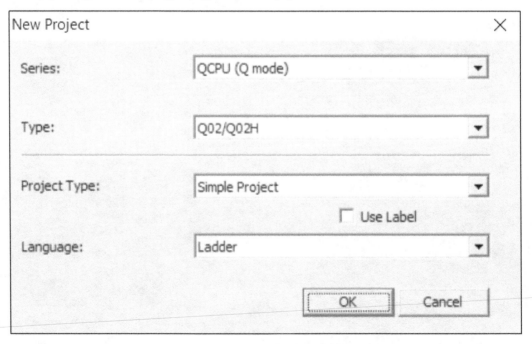

Series : Select the type of PLC. Select the series of PLC to be used here. For this program, the default setting is QCPU (Q mode).

Model : There are several models among the types of PLCs, depending on their program capacity and functions. Select the model of PLC you will use. In this programming, Q02/Q02H will be used.

Project Type : You can choose between "Simple Project" and "Structured Project. However, since we will not be dealing with a complex program in this study, we will use the default setting of "Simple Project. As for "Use labels," since label programming (defining signals/data used in a program as labels (variables)) is not used this time, there is no need to check this box.

Programming Language : You can choose between "Ladder" and "SFC". The default setting is "ladder".

A new project is created.

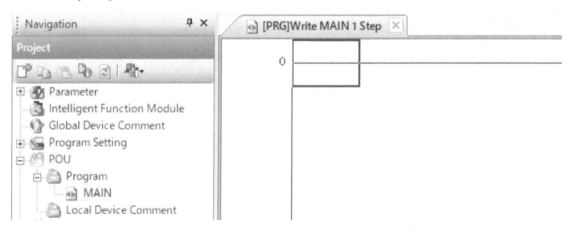

4-2 Saving a project.

1. select the "Project" tab in the upper left corner ⇒"Save As...". Select "Save " if you wish to overwrite the file.

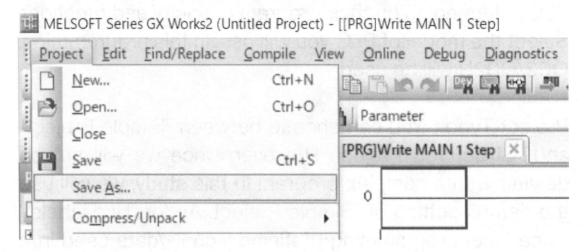

2. Name the file and press "Save"(保存).

4-3 Open a saved project.

1. Open GX Works2.

2. Select the "Project" tab in the upper left corner ⇒ "Open".

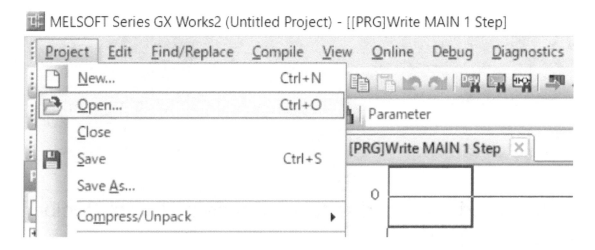

3. Select the saved file and press "Open"(開く).

4-4 Preparation before programming.

1. check if "View" tab ⇒"Toolbar" ⇒"Ladder" is checked.
If not, select it and check the box.

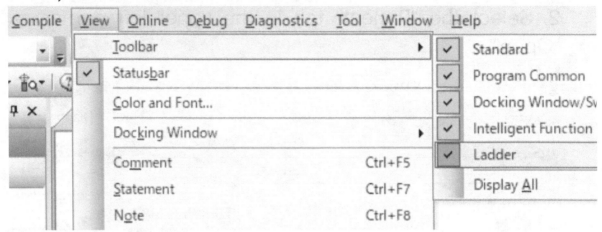

2. Ifchecked, a toolbarlike this will appear. Use this toolbar when creating ladders.

3. Select "Tool" tab ⇒"Options".

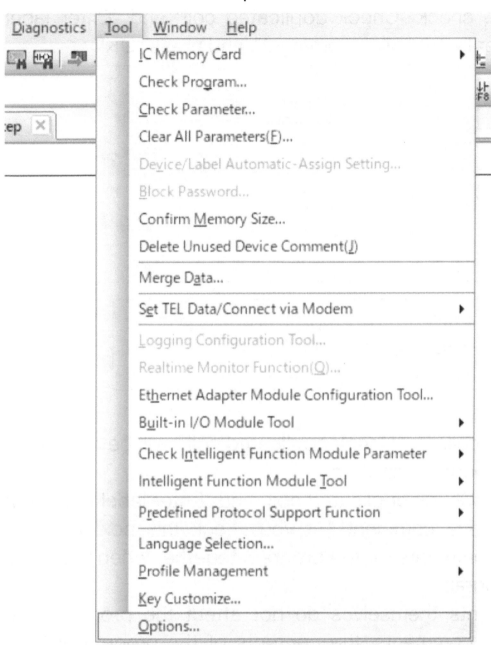

4. Select "Program Editor" ⇒ "Ladder/SFC" ⇒ "Enter ladder", check "Check doplicated coil" and "Enter label comment and device comment", and press "OK".

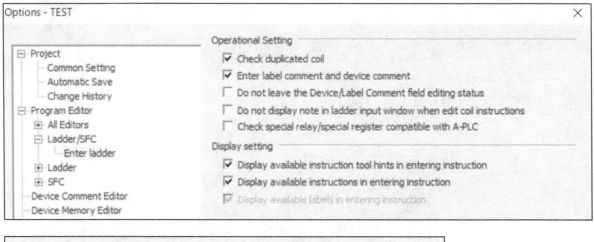

Double coils in " Check doplicated coil " will be discussed in subsequent chapters.

As for " Check doplicated coil" and "Enter label comment and device comment " if you check this box, you can enter comments for the program together when you write the program.

Comments themselves do not affect the program, but without comments, the contents of the program will be completely unknown except to the person who created it, so please make sure to comment on the program.

In addition to continuing to type, there is another way to enter a comment from the "Global Device Comments" menu in the navigation, but this is the way we will explain it this time.

4-5 Creation of program to check operation of switch lamp screen.

Nowthat we are ready to create the PLC program, we will now create the PLC program to check the operation of the switch lamp screen.

```
     X0
0   ─┤ ├────────────────────────────────(Y0    )
    Switch                                Lamp

     Y0
2   ─┤ ├──────────────────────────[INCP    D0    ]
    Lamp                                   Lamp
                                          lighting
                                          times
```

First, press F2 on the keyboard to enter write mode. Now let's start with the switch (X0). Press F5 on the keyboard or toolbar at the cursor position, enter "X0" in the Circuit input field, and then press "OK. Then, enter "switch" in the "Comment" field and press "OK.

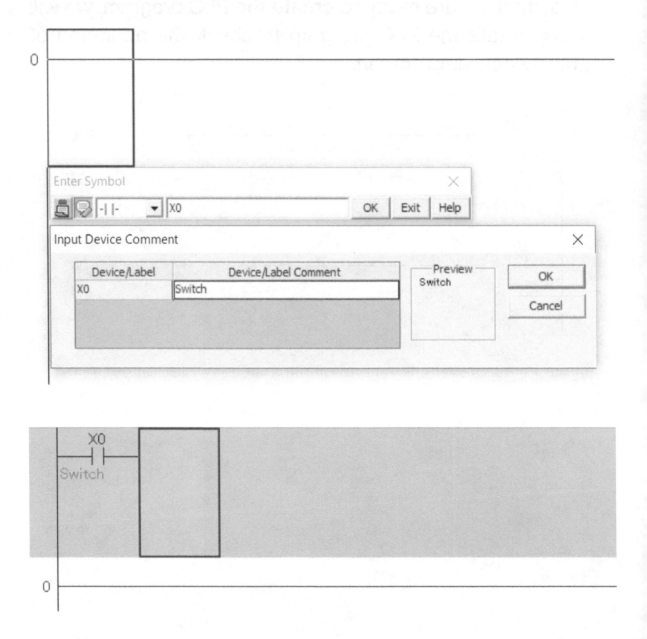

2. Next, create a lamp (Y0). Press F7 on the keyboard or toolbar at the cursor position, enter "Y0" in the "Circuit" field, and press "OK". Then, enter "Lamp" in the "Comment" field and press "OK.

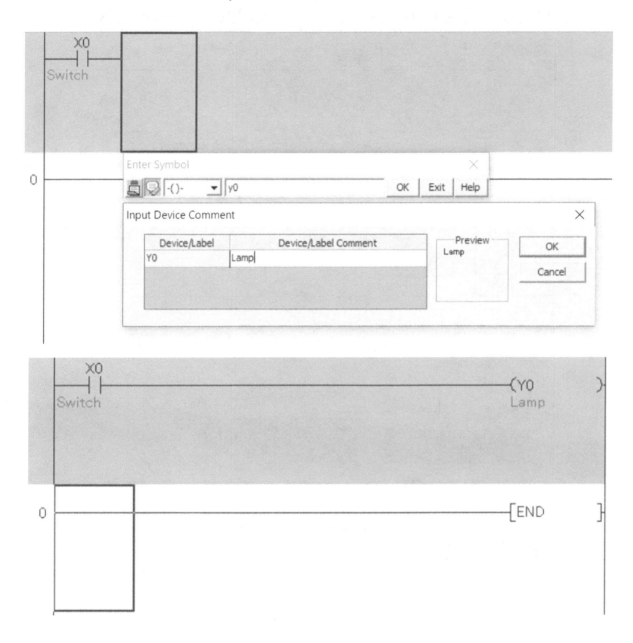

3. Create a circuit in which the lamp (Y0) is ON and +1 is put into the lamp lighting frequency (D0). First, press F5 on the keyboard or toolbar at the cursor position, enter "Y0" in the circuit input, and then press "OK". "Lamp" is entered in the comment input, and then press "OK".

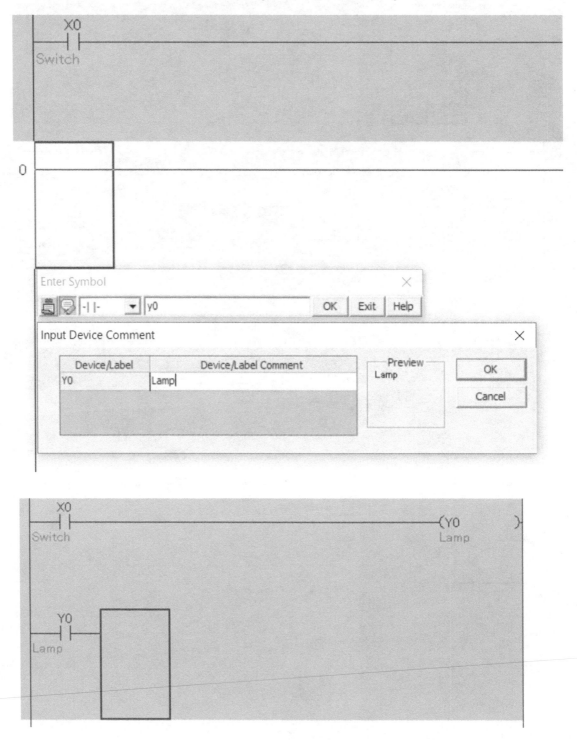

4. Press F8 on the keyboard or toolbar at the cursor position, enter "INCP D0" in the Circuit Input field, and press "OK. Then, enter "Lamp Lighting Count" in the "Comment" field and press "OK. INCP D0" is an instruction word that adds +1 to the value of D0 each time the condition is turned on.

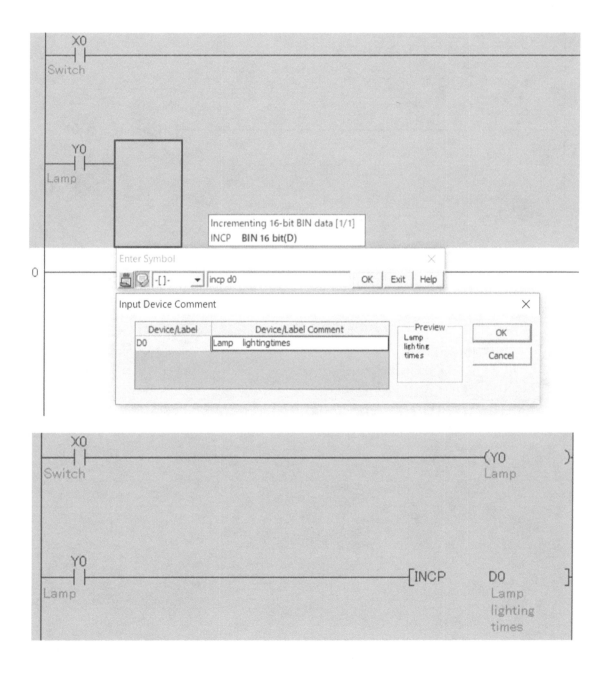

5. Press F4 on the keyboard to convert and the program creation is complete. (If you cannot convert4-10(If you cannot convert, see 4-10.) Also, do not forget to save the file.

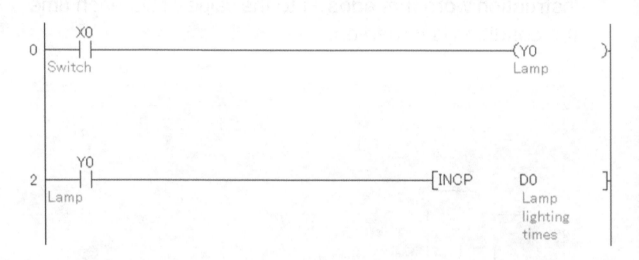

4-6 Start of PLC and touch panel simulation.

First, program settings for PC parameters. Click on Parameters in the navigation and double-click on PLC Parameter.

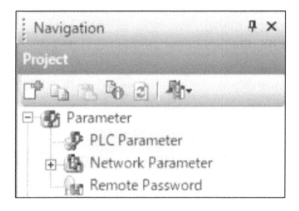

The parameter setting screen will appear. Press the "Program" tab, select the program "MAIN" on the left, and press the Insert button, after adding MAIN to the right side, press OK at the bottom right. The MAIN program will now be executed.

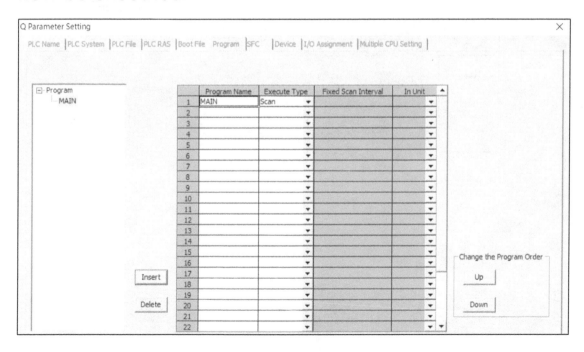

Next, start the simulation on the PLC program side.
Open the PLC program you just created and select
"Debug" tab ⇒"Start/Stop Simulation".

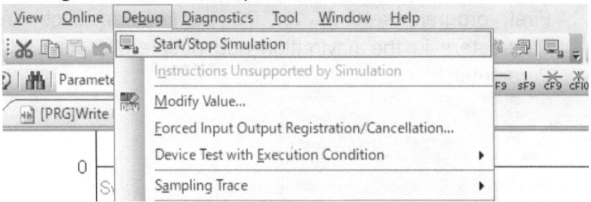

Simulation, but writing to the PLC will begin. When all is
complete, press "Close".

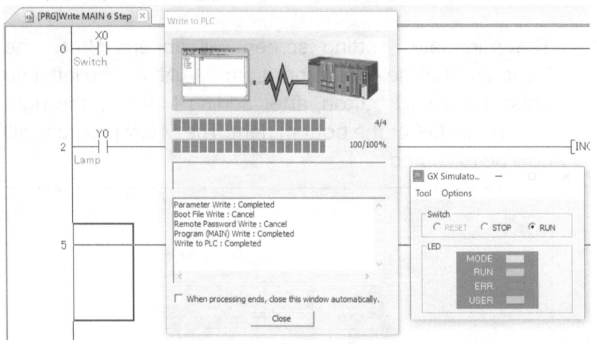

Now the PLC side is ready for simulation, and you can see that the number of times the D0 lamp is lit is 0.

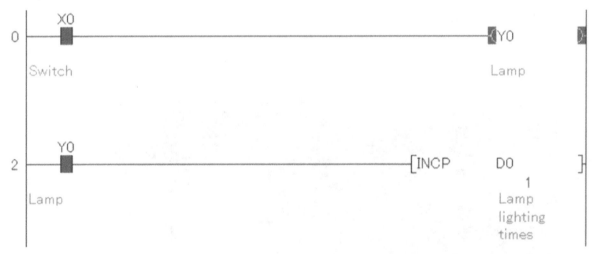

Also, when X0 is ON, Y0 is ON and D0 is 1. Thus, when X or Y is ON, it is blue. *Shift + Enter can be used to turn ON/OFF.

To end the simulation, select "Debug" tab ⇒"Start/Stop Simulation" again. Or close GX Works2 to end the simulation.

Next, start the simulation on the touch panel side.
Click "Tools" tab => Simulator => Activate.

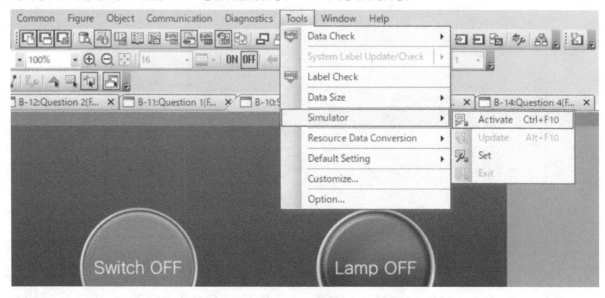

By the above operation, the simulation on the PLC side
and the simulation on the touch panel are linked, and the
simulator screen shown below is displayed.

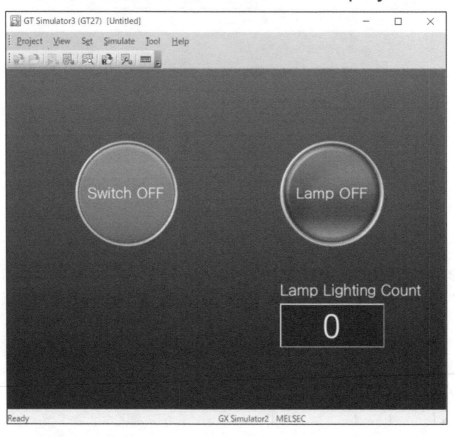

Whentheswitch is clicked onthe screen, the lamp is lit and 1 is placed in the lamp lighting count.

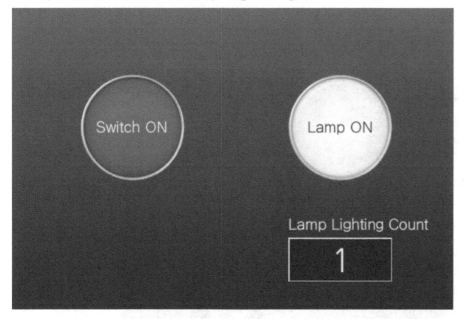

Also, each time the switch is pressed, the number of times the lamp is lit is +1.

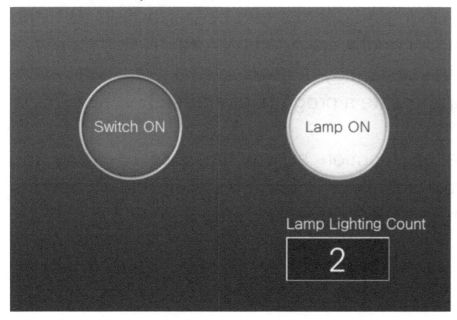

To exit the simulation, click the "Simulate" tab ⇒Stop, or press the "X" in the upper right corner.

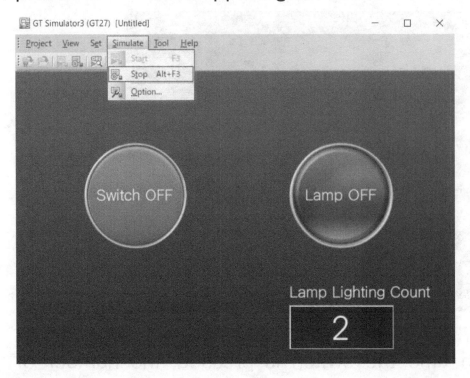

In this way, operation checks can be performed as if the PLC and touch panel were present. After all, it is more fun to operate the system by oneself, even if it is a simulation, than to simply create a program or screen.

This simulation function allows the user to verify the operation to some extent before installing PLC programs and screens in the actual equipment.

4-7 How to change the PLC program during simulation.

The simulation is in "monitor mode" when it is started. In this state, program changes cannot be implemented. If you want to change the PLC program during simulation, it is necessary to change to "monitor (write mode)" and the change procedure is described below.

Press "Online" tab ⇒"Monitor" ⇒"Monitor (Write mode)" or press Shift + F3 on the PC keyboard to bring up the following screen.

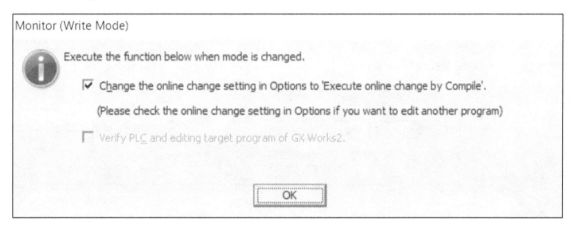

In this state, the program can be changed and written to the PLC on the simulation. Use this monitor (write mode) to add or modify the program if the simulation does not result in the expected behavior. Alternatively, you can exit the simulation mode once, add or modify the program, and then start the simulation mode.

To change the program in the monitor (write mode), for example, if you want to add +10 to the lamp lighting frequency (D0) when the lamp (Y0) is turned on, double-click [INCP D0], enter "+P K10 D0" in the circuit input, and then press "OK". Then press "OK" after entering "Lamp Lighting Count" in the comment input. *"+P K10 D0" is an imperative word that +10 isadded to the value of D0every time the condition is turned on.

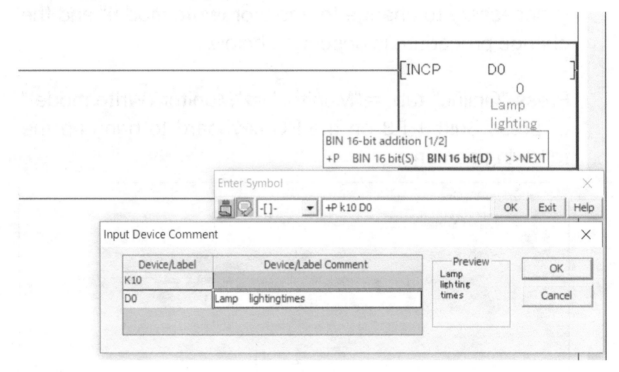

After changing the proglampress the "Compile" tab ⇒ "Online Program Change" or press Shift + F4 on the PC keyboard to get a message, then select "Yes" to write the program.

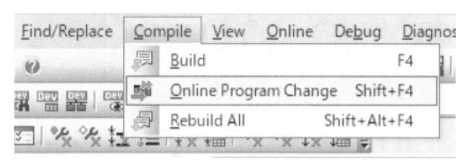

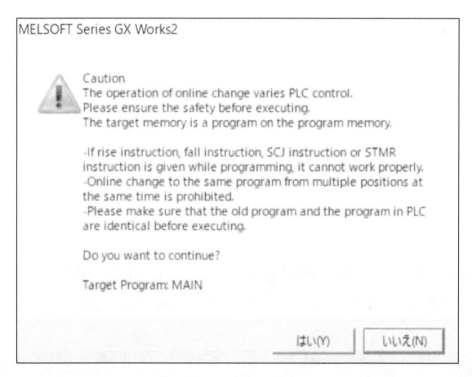

When the following message appears, writing during RUN is complete.

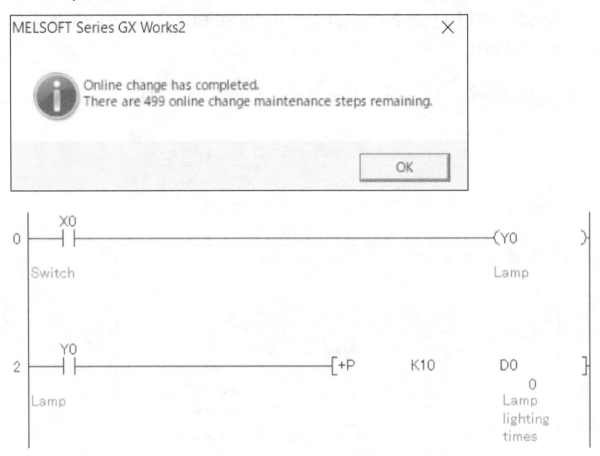

On thesimulation screen, verify that the number of times the lamp is lit increases by +10each time the switch is pressed.

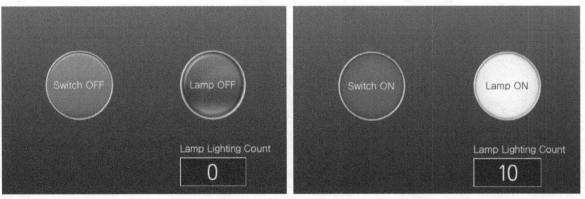

4-8 When a write message appears during RUN in program conversion.

After simulation of the PLC program is completed, the following message may appear when the program is changed and converted.

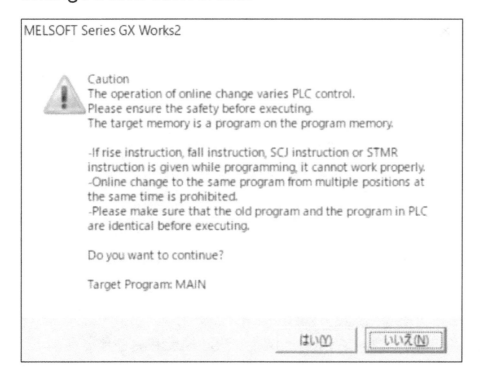

MELSOFT Series GX Works2

Caution
The operation of online change varies PLC control.
Please ensure the safety before executing.
The target memory is a program on the program memory.

-If rise instruction, fall instruction, SCJ instruction or STMR instruction is given while programming, it cannot work properly.
-Online change to the same program from multiple positions at the same time is prohibited.
-Please make sure that the old program and the program in PLC are identical before executing.

Do you want to continue?

Target Program: MAIN

はい(Y)   いいえ(N)

If this message appears, select "Write in RUN" from the "Tools" tab ⇒"Options" and uncheck "Execute online change by Compile".

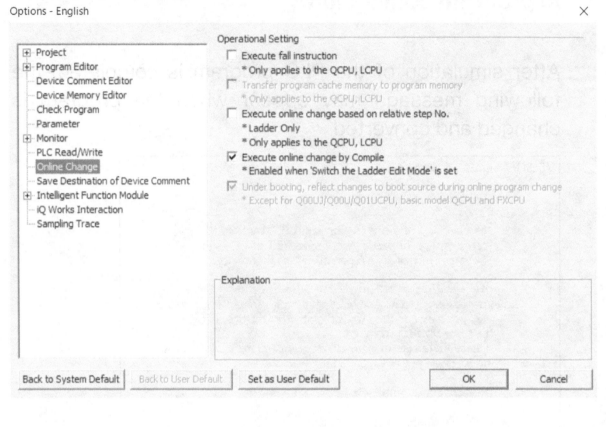

Execute online change by Compile
* Enabled when 'Switch the Ladder Edit Mode' is set

Also, if you cannot edit the proglampress F2 on the keyboard to enter write mode.

4-9 Checking the simulation operation of Exercises 1 to 4

Following the lamp switch screen, simulation of Exercises 1 through 4 will be used to check theoperation.

1. first, create the buttons for moving between the screens created in this case. On the Lamp Switch screen, select "Object" tab => Switch => Go to Screen Switch.

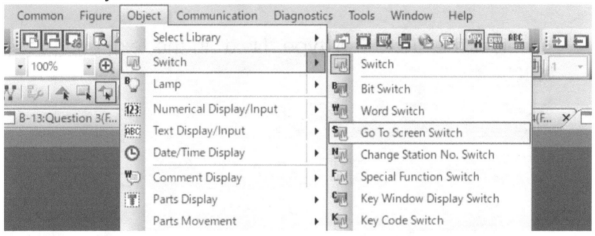

2. click on the screen to display the screen switching component.

3. Configure various settings for the screen switching component. Double-click the screen switching component to display the following screen. Enter "11" for the screen number.

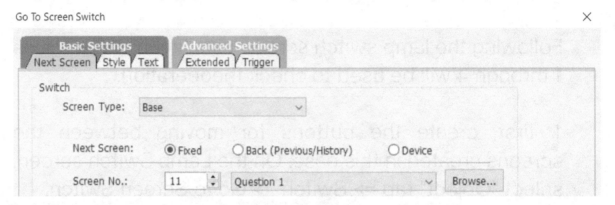

4. select the "Text" tab, type "Next" in the text, and press "OK".

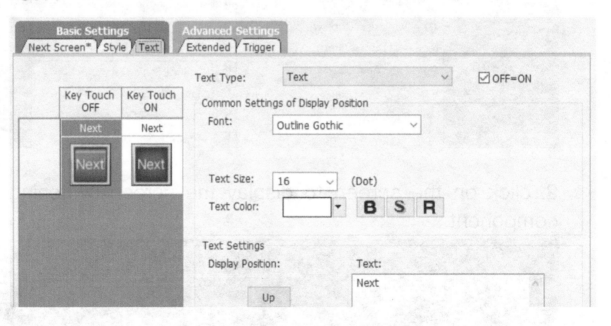

5. adjust the position and size of the part. After clicking on the screen switching component, adjust the position and size of the component in the lower left corner.
X:540 Y:430 Width:100 Height:50

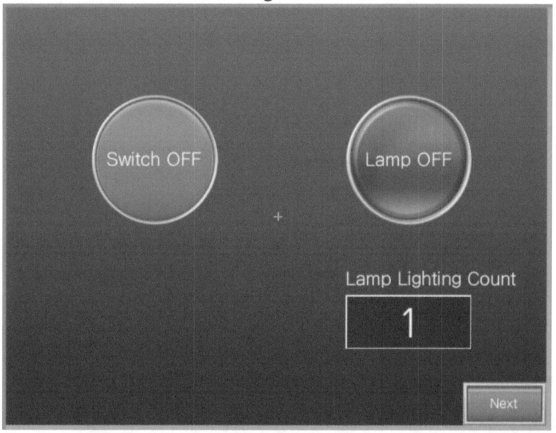

6. Follow the same procedure to create screen switching switches for screens 11 through 14. 6.

Screen 11
"Next" switch on screen 10.
Copy ⇒Paste
[Next Screen] Screen No. 10
[Text] Previous
Part location and size
X:0 Y:430 Width:100 Height:50

"Next" switch on screen 10.
Copy ⇒Paste
[Next Screen] Screen No. 12
[Text] Next
Part location and size
X:540 Y:430 Width:100 Height:50

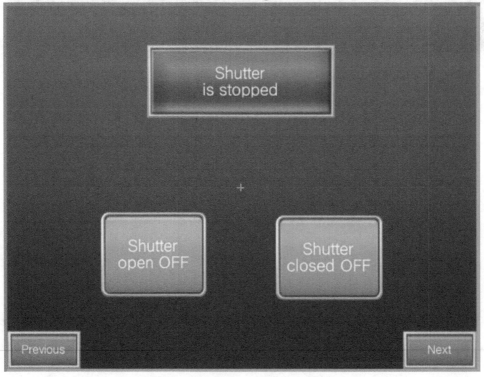

Screen 12

"Previous" switch on screen 11.

Copy ⇒Paste

[Next Screen] Screen No. 11

[Text] Previous

Part location and size

X:0 Y:430 Width:100 Height:50

"Next" switch on screen 11.

Copy ⇒Paste

[Next Screen] Screen No. 13

[Text] Next

Part location and size

X:540 Y:430 Width:100 Height:50

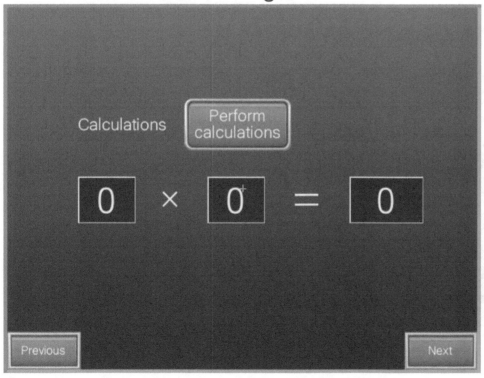

Screen 13

"Previous" switch on screen 12.

Copy ⇒Paste

[Next Screen] Screen No. 12

[Text] Previous

Part location and size

X:0 Y:430 Width:100 Height:50

"Next" switch on screen 12.

Copy ⇒Paste

[Next Screen] Screen No. 14

[Text] Next

Part location and size

X:540 Y:430 Width:100 Height:50

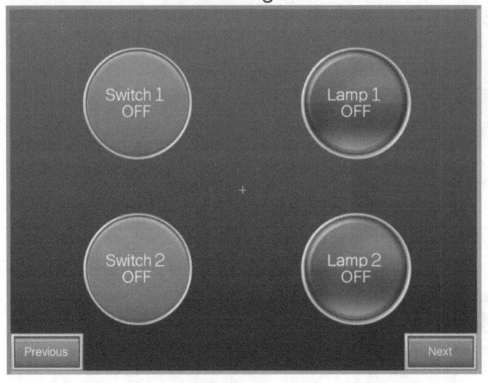

Screen 14

"Previous" switch on screen 13.

Copy ⇒Paste

Switching destination setting] Screen No. 13

[Text] Previous

Component location and size

X:0 Y:430 Width:100 Height:50

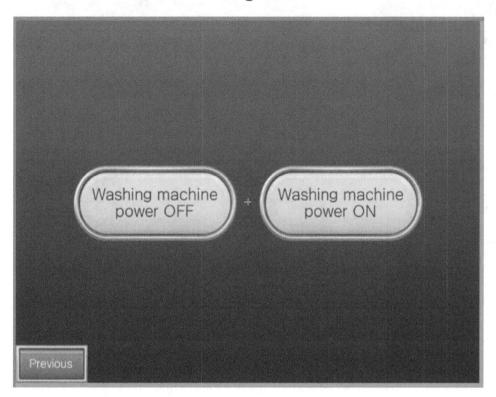

7. Check the created screen switching switch on the screen for correct operation.

Select "View" tab => Preview.

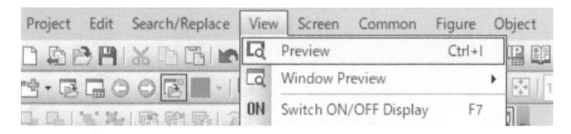

8. The screen preview screen will be displayed, so confirm that the screen moves correctly by pressing the "Next" or "Previous" switch. After confirming, press the "X" in the upper right corner to close the window.

9. Next, create a PLC program for simulation for each problem. (Press F2 to enter write mode.) First, to add the PLC program for problem 1, insert a line under the program for the lamp switch screen. Edit" tab => Insert Row or Shift+Insert the keyboard will insert a line, please insert two lines.

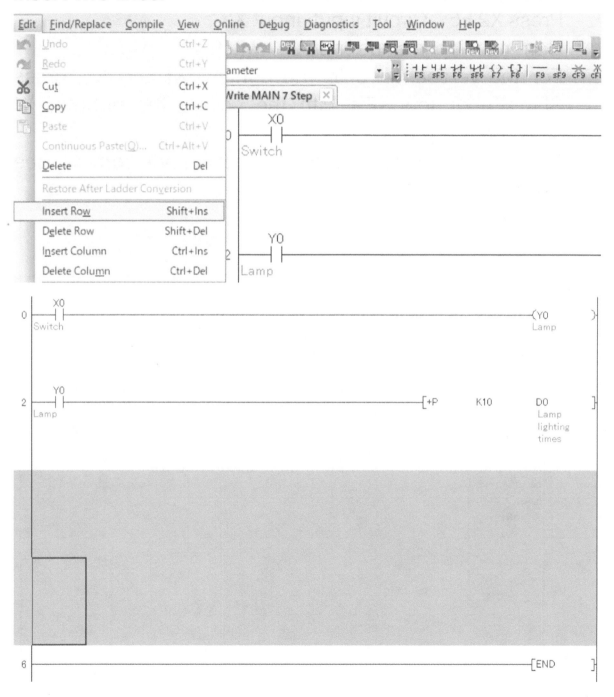

10. follow the steps below to create the PLC program for Problem 1.

Press X1:F5. Circuit input: Input X1

Comment input: Shutter open switch

Y1: Press F7. Circuit input: Input Y1

Comment input: Shutter operating lamp

Press X2:F5 Circuit input: Input X2

Comment input: Shutter close switch

For the vertical lines to the right of X1 and X2, press Shift+F9 or tool sF9 with the cell to the right of X1 selected and press "OK".

Finally, pressF4 to convert. (If you cannot convert, see 4-10)

11. Now that we are ready to check the operation of Problem 1 by simulation, we can start the simulation of GX Works2 and the simulator of GT Designer3.

Go to Screen 11 (Problem 1) and verify that the lamp goes to "Shutter is in operation" whenthe "Shutter open switch" or "Shutter close switch" is pressed.

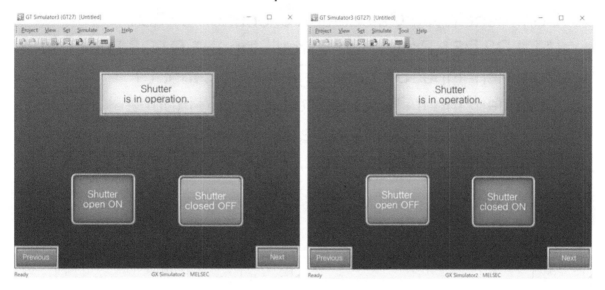

12. If the simulation does not work as expected, check the PLC program, devices on the screen, etc. for errors.

If the correction point is in the PLC program, use the monitor (write mode) to correct the program, or stop the simulation once, correct it, and start the simulation again.

If the correction is on the screen, after correcting the screen, press "Tools" tab ⇒Simulator ⇒Update, or close the simulator and start it upagain.

13. Once the simulation for problem 1 is finished, exit the PLC program andscreen simulation and create the PLC programs for simulation of problems 2, 3, and 4.

Problem 2: PLC program for simulation
Press X3:F5   Circuit Input: Input X3
Comment input: Calculation execution switch
[* D1 D2 D3]: Press F8   Circuit Input: * D1 D2 D3
Comment input D1: One-digit numeric input 1
Comment input D2: One-digit numeric input 2
Comment input D3: Calculation results
Finally, press F4 to convert.

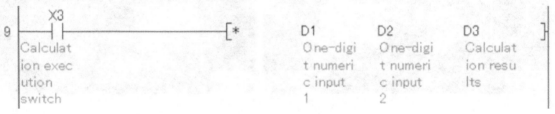

Issue 2 Simulation
Go to Screen 12 (Problem 2), enter the numbers in the before and after rows of the calculations, and then press the Perform Calculation switch to confirm that the result of the calculations is correct.

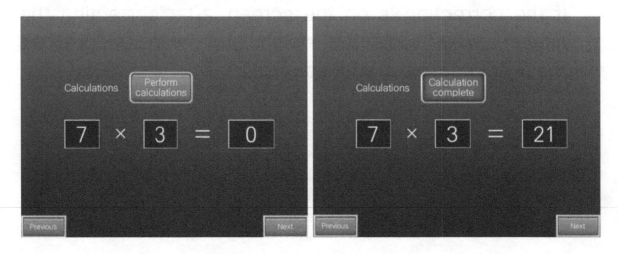

Problem 3: PLC program for simulation

X10: Press F5   Circuit input: Input X10

Comment input: Switch 1   Y10: Press F7

Circuit input: Input Y10   Comment input: Lamp 1

X11: Press F5   Circuit input: Input X11

Comment input: Switch 2   Y11: Press F7.

Circuit input: Input Y11   Comment input: Lamp 2

Finally, press F4 to convert.

```
      X10
13   ─┤ ├─────────────────────────────────(Y10        )─
     Switch 1                              Lamp 1


      X11
15   ─┤ ├─────────────────────────────────(Y11        )─
     Switch 2                              Lamp 2
```

Problem 3 Simulation

Go to Screen 13 (Problem 3) and verify that switch 1 turnslamp 1 ON for as long as it is pressed, switch 2 remains ON and lamp 2 turns ON when pressed while switch 2 is OFF, and switch 2 remains OFF and lamp 2 turns OFF when pressed while switch 2 is ON.

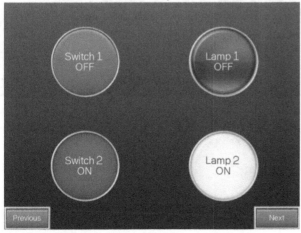

Problem 4: PLC program for simulation
X20: Press F5   Circuit input: Input X20
Comment input: Washing machine power on switch.
Y20: Press F7   Circuit input: Input Y20
Comment input: Washing machine power signal.
Finally, press F4 to convert.

```
       X20
17   ┤ ├                                                    ─(Y20      )─

    Washing                                               Washing
    machine                                               machine
    power on                                              power
     switch                                               signal
```

Issue 4 Simulation
Go to Screen 14 (Problem 4) and confirm that pressing
the washer power on button turns the washer power on
and pressing the washer power off button turns the
washer power on button off.

Chapter5 Frequently Used Functions

5-1 Copy/Paste/Undo/Redo shortcuts

All these operations can be selected from the "Edit" tab, but please learn to use the shortcuts, as they will greatly speed up the creation of your screen. (The operations are the same as in Excel.)

Copy: Ctrl+C
⇒Copyparts and shapes

Paste: Ctrl+V
⇒Pastecopied parts and figures.

Undo: Ctrl+Z
⇒Reverts one wrong operation, etc. to the previous one.

Redo: Ctrl+Y
⇒Redo the returned operation and proceed one step further.

5-2 Device display

You can double-click on a part you created on the screen to see its devices, but if there are many parts on the screen, it is tedious to double-click on each one to see them. In that case, you can select the "View" tab ⇒ Display Items ⇒Device to display the devices for each part.

If you wish to turn off the device display, follow the same procedure to uncheck the box.

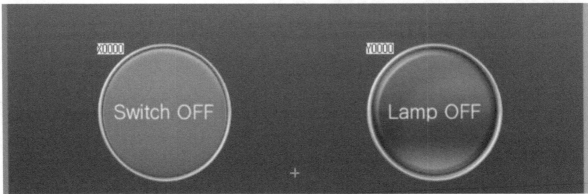

5-3 Device Search

If you have created multiple screens and want to search for the device you are using, click on the "Search/Replace" tab ⇒"Device Search" or press Ctrl + F keyboard to display the device search window.

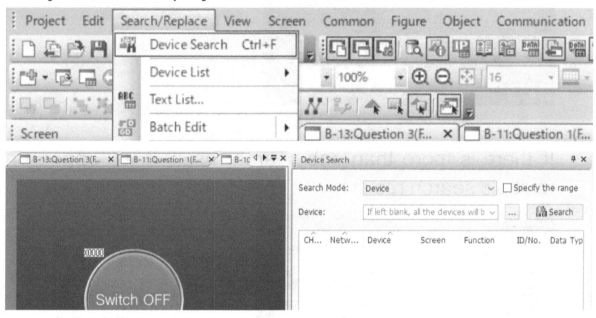

After entering the device, you wish to search for (X0010 in this case) in the device search window, press the search button and X0010 will appear in the search results.

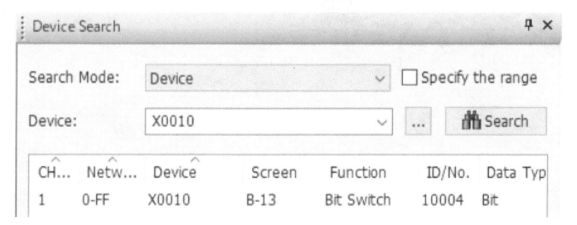

Then click on X0010 in the search results to go to the screen with the button.

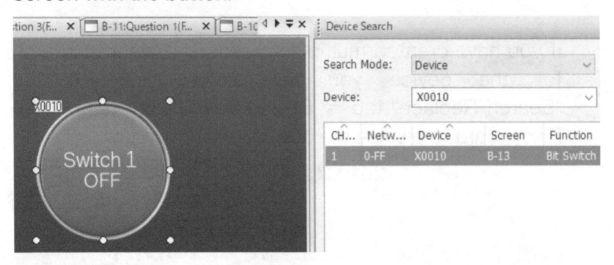

If there is more than one search result, it will appear in multiple search results.

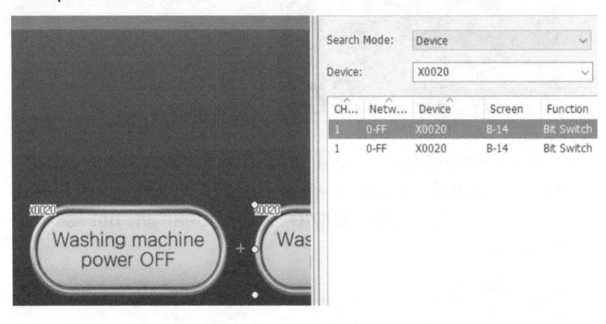

5-4 Grid

Displaying the grid makes it easier to adjust the position of each component.
The grid is selected from the "View" tab ⇒Grid.

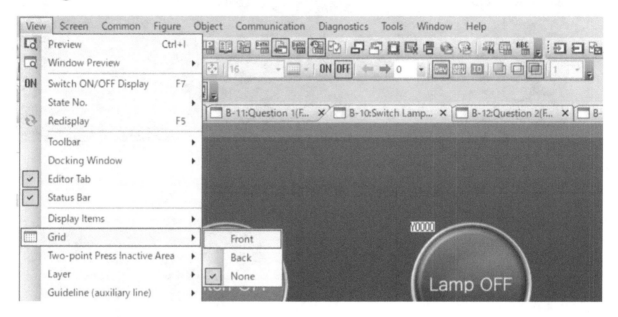

If you select the front side, the grid is also displayed on the part.

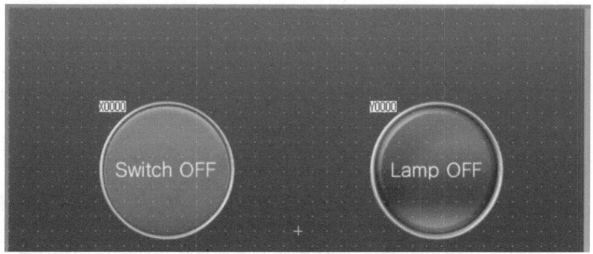

If you choose the back side, no grid will be displayed on the part.

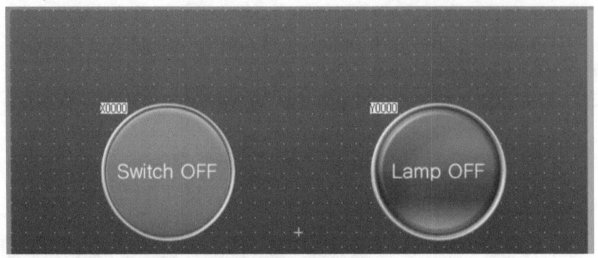

You can also change the grid spacing.

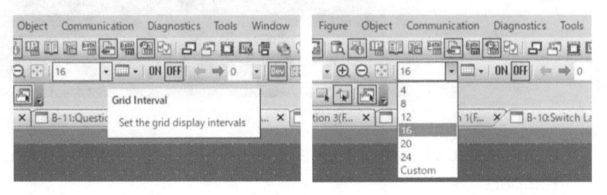

When the grid spacing is changed from the standard 16 to 8

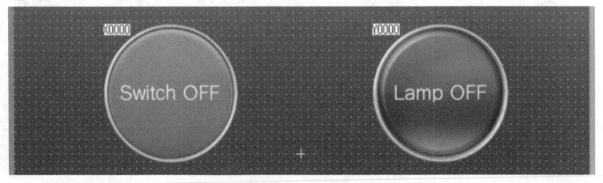

5-5 Fine adjustment of part position

After selecting a part, you can move it with the cursor, but if you want to fine-tune the position of the part, hold down Alt on the keyboard and move the cursor up, down, left, or right. (You can also use XY in the lower left corner.)
You can also change the amount of movement with the cursor.

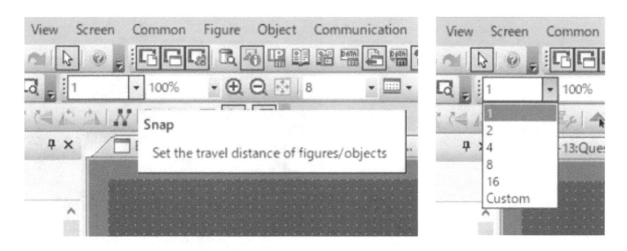

For example, if you set the amount of movement to 8, pressing the cursor once will move the cursor by 8.

5-6 Zoom

If you want to enlarge the screen and make fine position adjustments, use the "View" tab => Zoom to select the desired percentage to display and adjust the position.

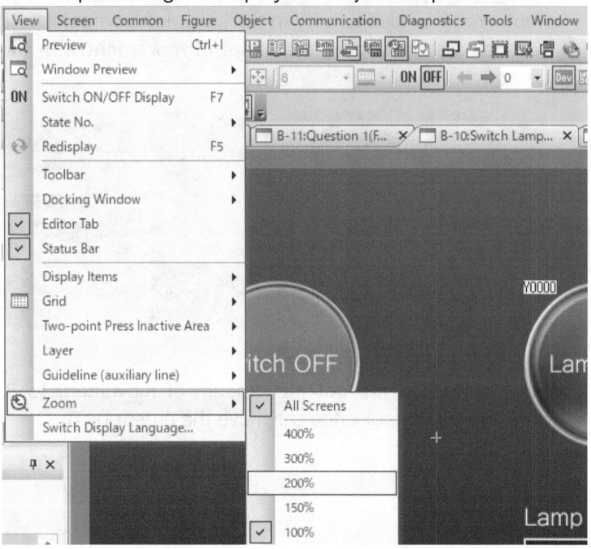

At 200% zoom

Zoom magnification can also be changed from the following locations.

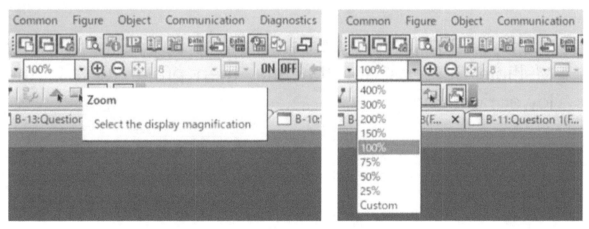

5-7 Continuous copy

Continuous copying is a useful feature when creating multiple copies of the same part. We will now explain how to use continuous copying while creating the screen below.

1. First, double-click on "New" under the Base Screen folder on the screen.

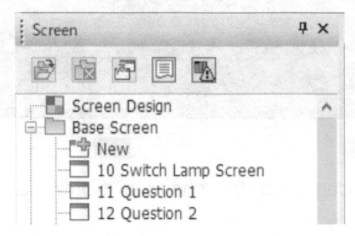

2. enter "20" for the screen number and "Switch Lamp Multiple Screens" for the title, then press "OK" in the lower right corner.

Screen Property

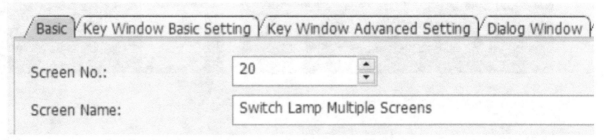

3. first create switch 1 and lamp 1.

Switch 1 (X0)

Switch 1 on screen 13.

Copy ⇒Paste

Basic Settings (Devices)

Device: X0000

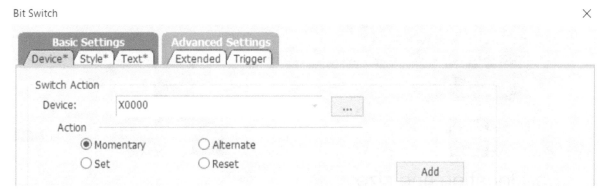

[Basic settings (text)]

OFF Text size:16　ON Text size:16

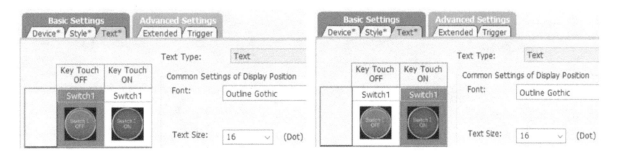

Part location and size

(Displayed in the lower left corner after clicking on a part)

X:130 Y:20 Width:100 Height:100

Lamp 1 (Y0)
Lamp 1 on screen 13.
Copy ⇒Paste
Basic Settings (Device/Style)
Device: Y0000

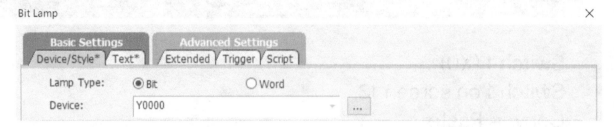

[Basic settings (text)]
OFF Text size:16　ON Text size:16

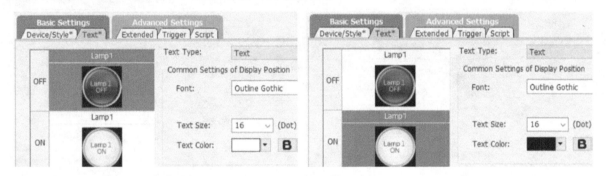

Part location and size
(Displayed in the lower left corner after clicking on a part)
X:400 Y:20 Width:100 Height:100

4. Select Switch 1 and Lamp 1 by either framing them while holding down the left mouse button or selectingbothwhile holding down the Shift button, resulting in the screen shown below.

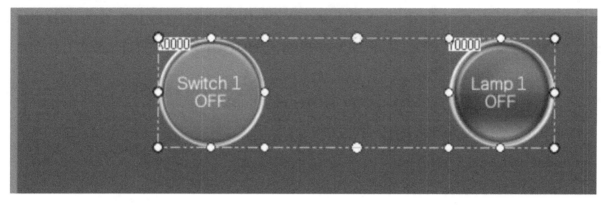

5. select "Edit" tab => Consecutive Copy or right-click on the part => Consecutive Copy.

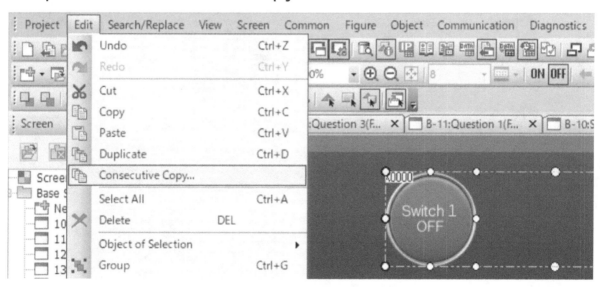

6. The continuous copy setting screen will appear. Since we will be creating four switches and four lamps vertically (Y direction), set the number in the Y direction to 4 and the spacing to 15 dots. The range/direction can be left as is. We also want the devices of the part to be X0, X1, X2, X3 or Y0, Y1, Y2, Y3, so we make them increment targets.

Device No: Check

Target device: Monitor/lamp devices

Increment setting: All、1

and press OK.

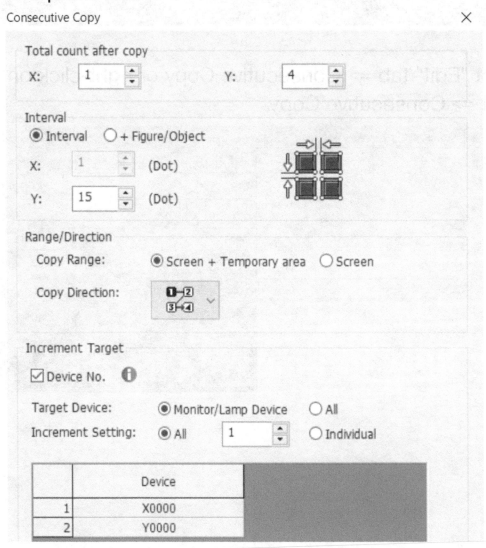

8. After the continuous copying is performed, the screen below will be created. All that remains is to correct the ON/OFF text for switches 2 to 4 and lamps 2 to 4, and the process is complete.

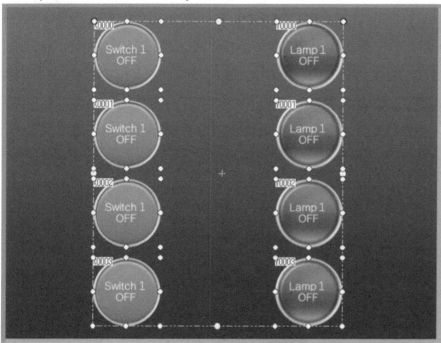

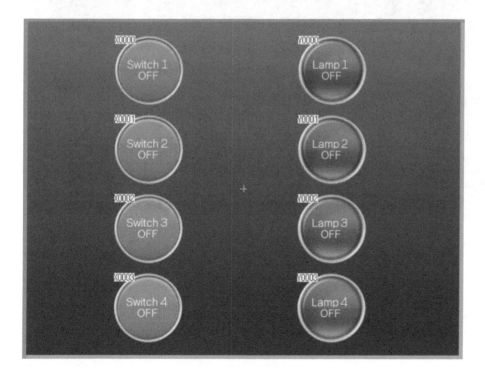

5-8 Alignment

Alignment is used to neatly align multiple parts.
After selecting each component of the screen, you just created which hasfour lamps and four switches each, move them around while left clicking to replace them with clutter. We will explain how to use alignment as we adjust this screen.

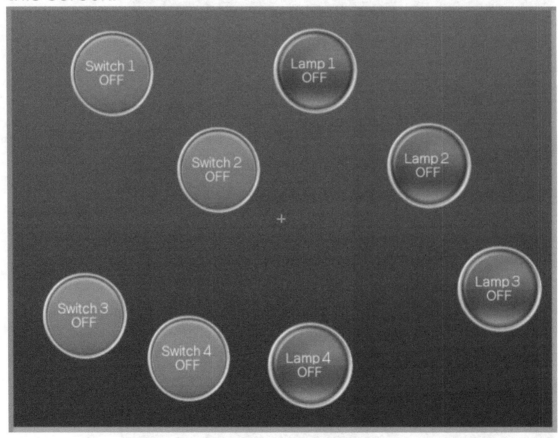

1. First, align the switches by left-clicking and circling or Shift-selecting the four switches in the screen below.

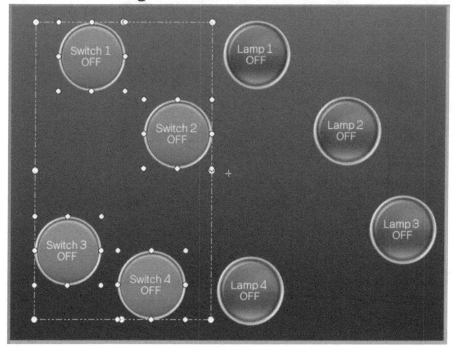

2. With the four parts selected, select "Edit" tab ⇒Align ⇒Center(Horizontal), or right-click on the part ⇒Align ⇒ Center(Horizontal) to center theleft and right sides of the parts.

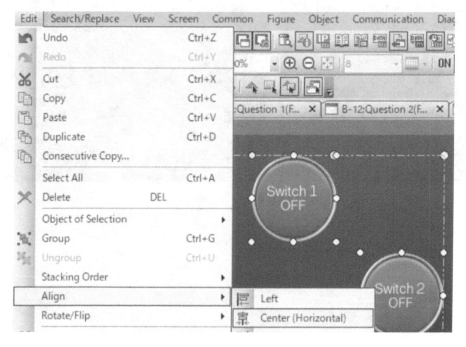

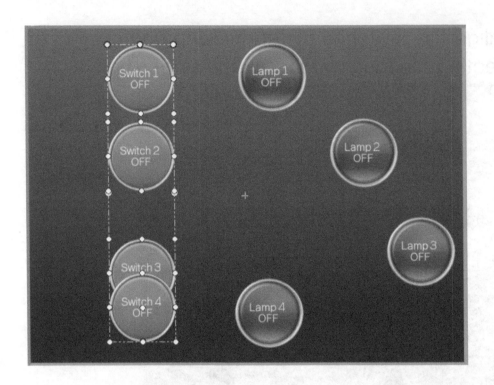

Left alignment is based on the left-most part, and right alignment is based on theright-most part.

For left alignment.   In case of right aligned.

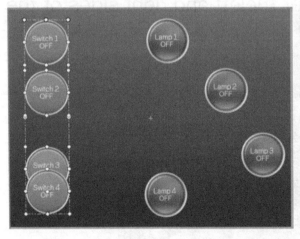
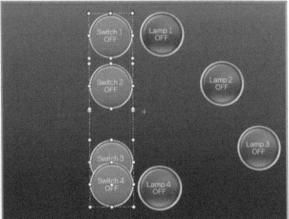

3. with the same four parts selected, select "Edit" tab ⇒ Align ⇒Align Lengthways or right-click on the part ⇒ Align ⇒ Align Lengthways to align the parts evenly vertically.

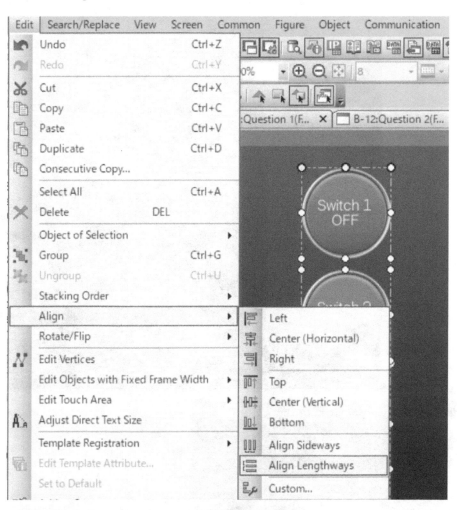

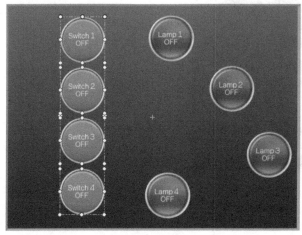

153

4. next, align the lamps. Align the lamps in the center using the same procedure as for switches.

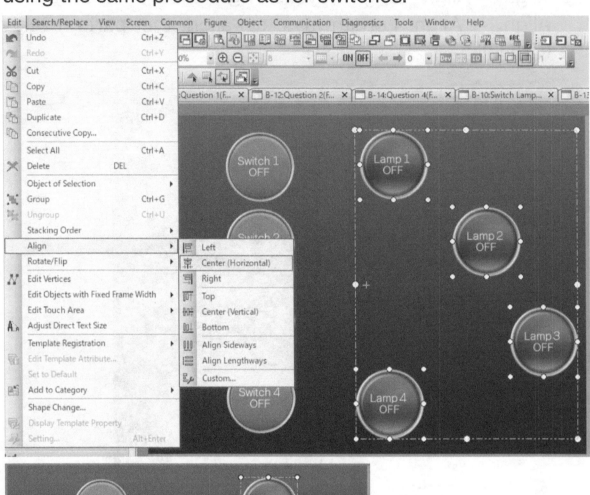

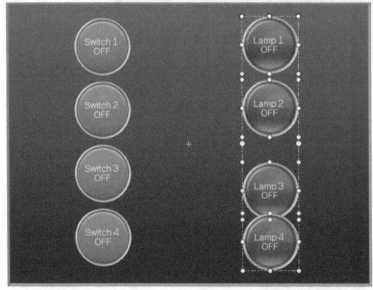

5. Since Lamp 1 is higher than Switch 1, select Switch 1 and Lamp 1, then select "Edit" tab ⇒ Align ⇒ Bottom, or right-click on the part ⇒ Align ⇒ Bottom to align the part with the switch at the bottom.

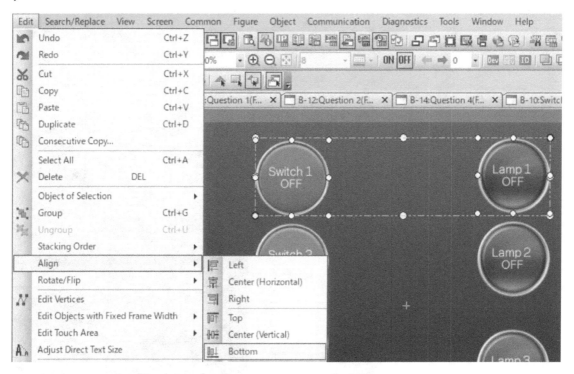

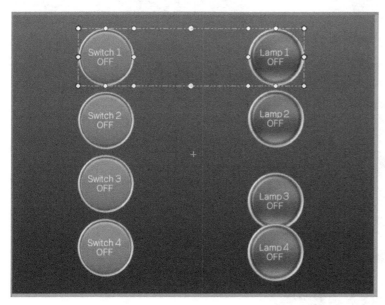

6. Lamp 4 is lower than Switch 4, so select Switch 4 and Lamp 4, then select the "Edit" tab and choose ⇒ Align ⇒ Top, or Right-click on the part and choose Align ⇒ Top to align the part with the switch at the top.

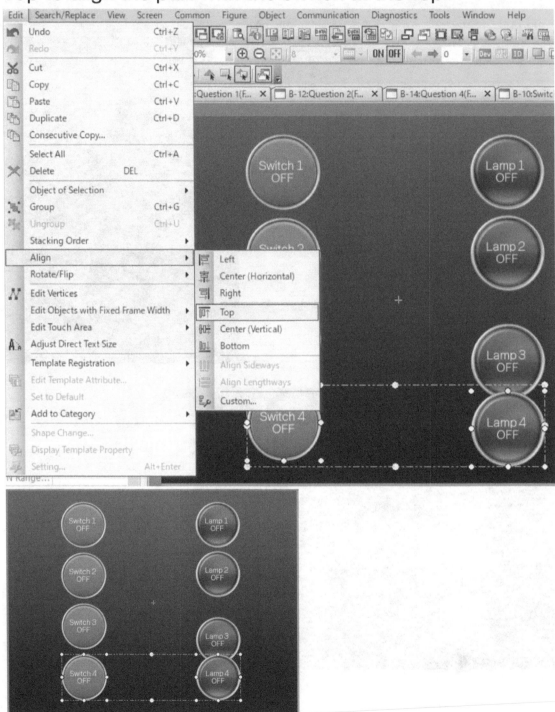

7. After selecting the four switches, align them evenly vertically to complete the alignment.

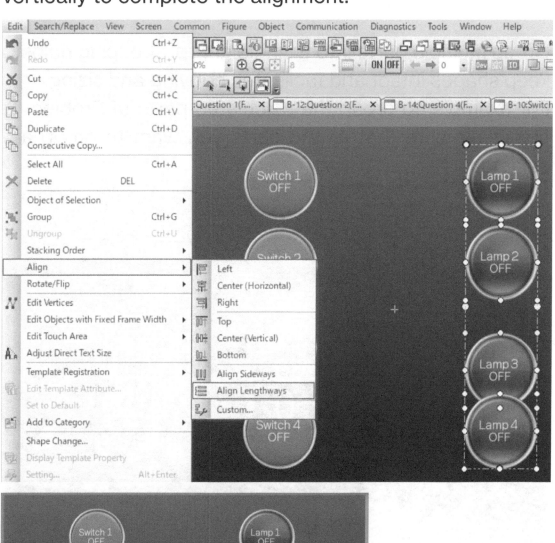

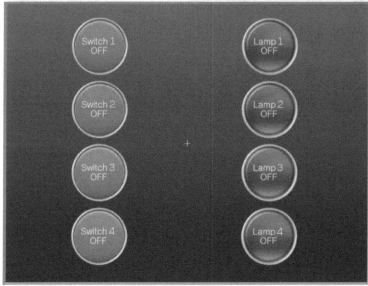

5-9 Grouping

Group multiple parts into a single group. Grouping helpsto keep shapes intact and facilitates copying and sizing.
For example, to group the calculated parts of Problem 2, first make sure that all the parts you want to group are selected.

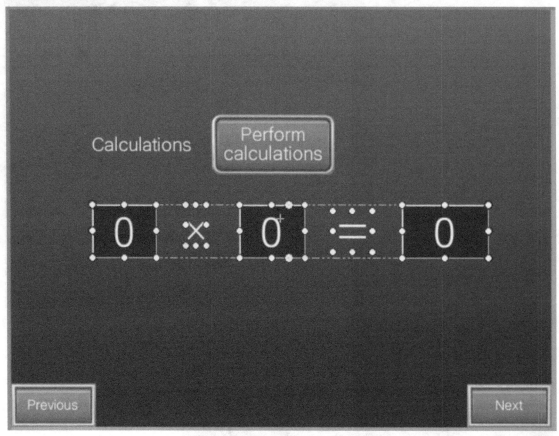

Choose "Edit" tab => Group or right click => Group. (Ctrl+G is also OK)

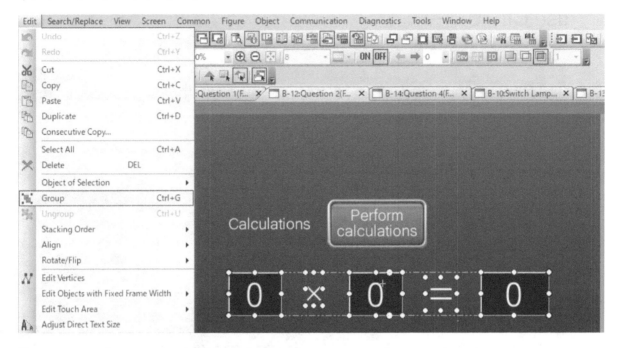

By grouping, when moving, you can also select one part and move it, and all the grouped parts will move together.

Also, after grouping, click and hold left-click and pull on the white circle in the frame to adjust the size of all grouped parts.

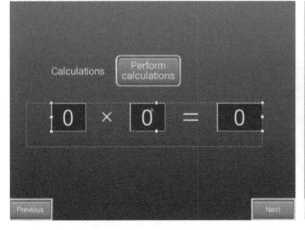 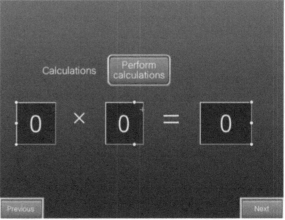

To ungroup, select the grouped parts, then select "Edit" tab ⇒Ungroup or right-click ⇒Ungroup. (You can also use Ctrl+U)

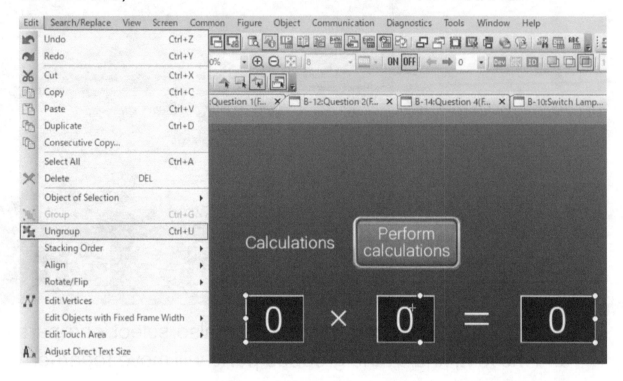

5-10 Batch change of strings

The text ofa shape or part can be changed after double-clicking on the shape or part.

However, as the number of screens increases, there will be more and more situations where you will want to change the strings ofmultiple components at once. In such cases, batch string changes are useful.

For example, the following is a procedure for copying and pasting the switch lamp multiple screens created earlier, creating a new screen with screen number 21 and title: Button Lamp Multiple Screens, and changingall "switch" strings to"button" strings.

1. right-click on screen number 20 ⇒select Copy, then right-click on screen number 20 ⇒select Paste.

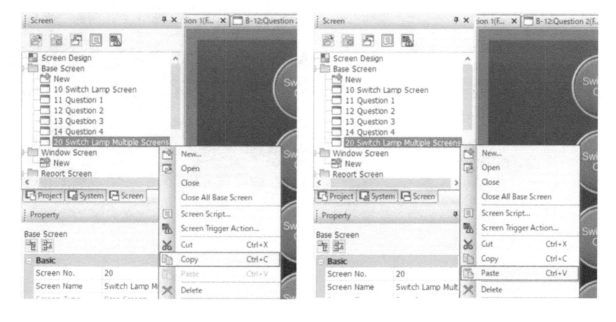

2. The screen properties will be displayed. Enter screen number: 21, title: Button Lamp Multiple Screens, and then press "OK".

Screen Property

| Basic | Key Window Basic Setting | Key Window Advanced Setting | Dialog Window | Option Selection Window |

Screen No.: 21

Screen Name: Button Lamp Multiple Screens

3. Double-click on screen number 21 to display it, then select the "Search/Replace" tab ⇒Data Browser.

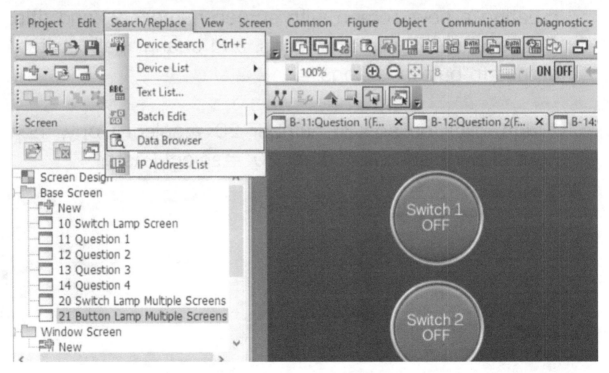

4. The Data Browser will appear. Click on the "Text" area so that switches 1 through 4 are lined up.

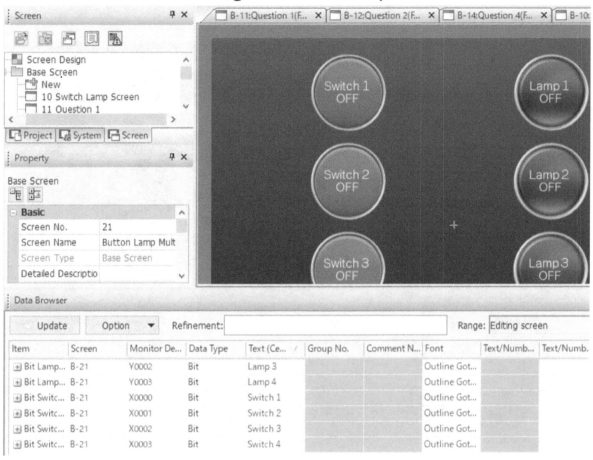

Item	Screen	Monitor De...	Data Type	Text (Ce...	Group No.	Comment N...	Font	Text/Numb...	Text/Numb.
⊞ Bit Lamp...	B-21	Y0002	Bit	Lamp 3			Outline Got...		
⊞ Bit Lamp...	B-21	Y0003	Bit	Lamp 4			Outline Got...		
⊞ Bit Switc...	B-21	X0000	Bit	Switch 1			Outline Got...		
⊞ Bit Switc...	B-21	X0001	Bit	Switch 2			Outline Got...		
⊞ Bit Switc...	B-21	X0002	Bit	Switch 3			Outline Got...		
⊞ Bit Switc...	B-21	X0003	Bit	Switch 4			Outline Got...		

5. after clicking switch 1 in thedata browser, click switch 4 while holding down Shift on the keyboard, as shown in the screen below.

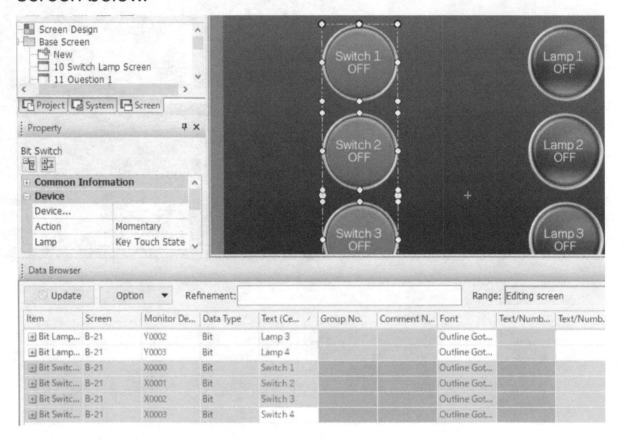

6. select Data Browser Option ⇒ Text Batch Edit.

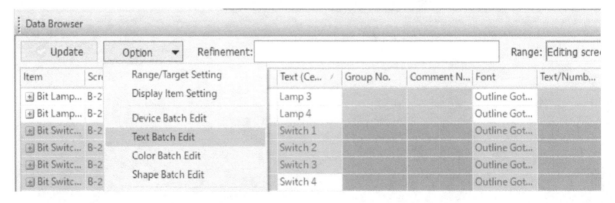

7. Enter "switch" for the string to be changed and "button" for the string after the change, then press the "Replace" button.

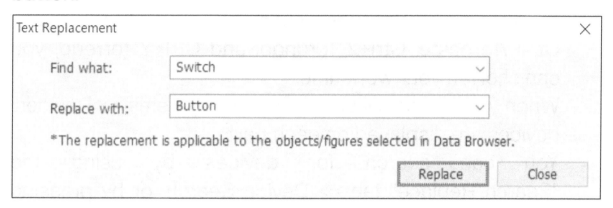

5-1 to 5-10 Summary

If you remember the shortcuts Ctrl+C for copy, Ctrl+Vforpaste, Ctrl+Z forundo, and Ctrl+Y forredo, you can shorten your work time.

When you select "View" tab => Display items => Devices, devices are displayed on each part.

You can search for devices by using the "Search/Replace" tab => Device Search, or by pressing Ctrl+F on your keyboard.

The grid view in the "View" tab ⇒Grid makes it easier to adjust the position of each component.

To fine-tune the position of a component, hold down Alt on the keyboard and use the up/down/left/right cursors to fine-tune the position.

The "View" tab ⇒Zoom allows you to zoom in and out on the screen.

Consecutive copies are useful when creating multiple copies of the same part.

Alignment is used to neatly align multiple parts.

Grouping allows you to group multiple parts into a single group. Grouping keeps shapes intact and facilitates copying and size adjustment.

Batch change ofstrings is used when you want to change the strings of multiple parts at once.

Exercise 5 (Alignment of Multiple Parts)

Create a new screen number 15, question 5, and copy =>
paste all theparts of the switch lamp multiple screens to
screen number 15, then use the grid and cursor
movement to display the device after moving it as shown
in the following screen.

Before change

After change

Exercise 5 Solution

1. After double-clicking on "New ", enter screen number: 15, title: Problem 5, and press "OK" to create a new screen.

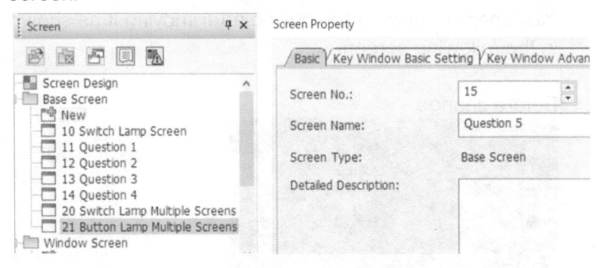

2. Screen 20: Copy and paste all the parts of the switch lamp multiple screens to screen 15.

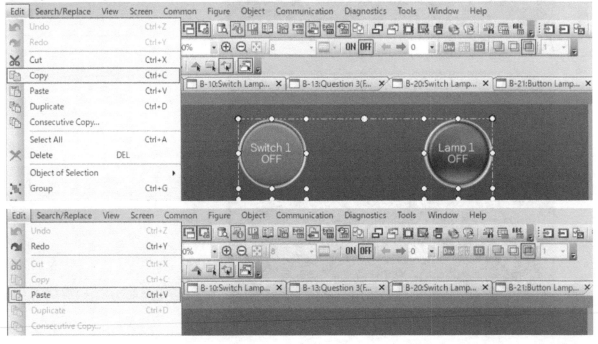

3. select "View" tab ⇒Grid ⇒Front to display the grid.

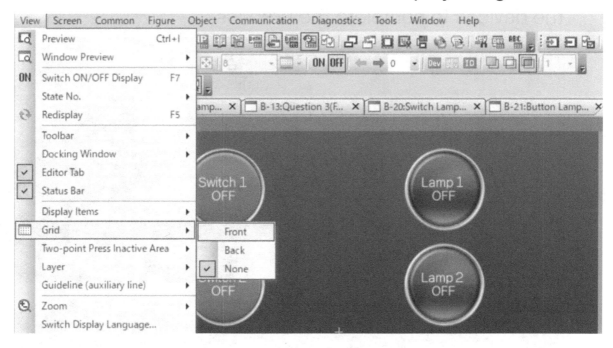

4. Click the part you want to move, then move the part while holding down the cursor or left mouse button to move it in accordance with the grid.

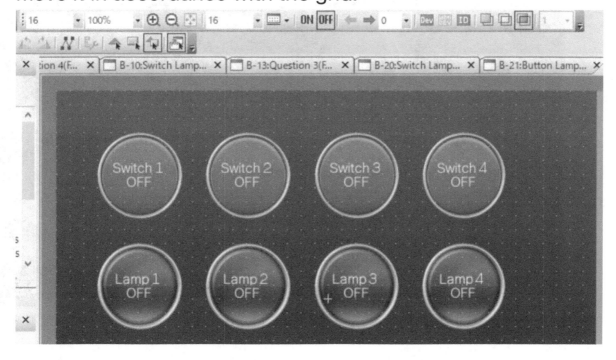

5. select "View" tab => Display Items => Device to display the devices to complete the process.

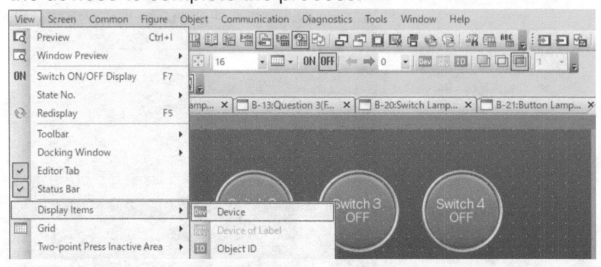

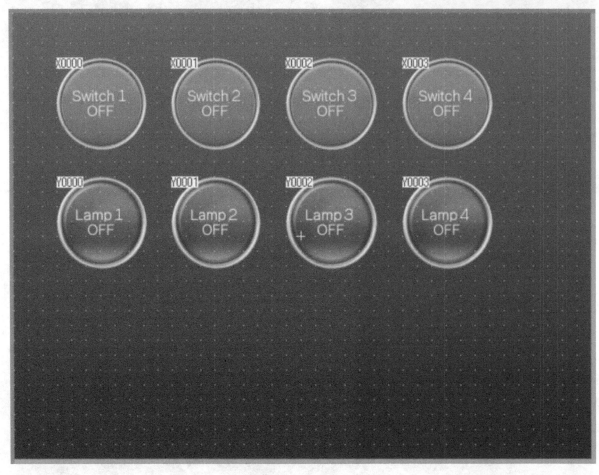

Exercise 6 (display of step history)

Create a new screen number 16, question 6, and after creating the title and 01-day step count display (D100) using the text of the shape and the numerical display of the object, create a screendisplaying the April step count history using continuous copy and batch change of strings. Devices used: D100 to D129
The 20th and 30th are not subject to batch string changes.

April step count history

01th	D100 2345	Steps	11th	D110 2345	Steps	21th	D120 2345	Steps
02th	D101 2345	Steps	12th	D111 2345	Steps	22th	D121 2345	Steps
03th	D102 2345	Steps	13th	D112 2345	Steps	23th	D122 2345	Steps
04th	D103 2345	Steps	14th	D113 2345	Steps	24th	D123 2345	Steps
05th	D104 2345	Steps	15th	D114 2345	Steps	25th	D124 2345	Steps
06th	D105 2345	Steps	16th	D115 2345	Steps	26th	D125 2345	Steps
07th	D106 2345	Steps	17th	D116 2345	Steps	27th	D126 2345	Steps
08th	D107 2345	Steps	18th	D117 2345	Steps	28th	D127 2345	Steps
09th	D108 2345	Steps	19th	D118 2345	Steps	29th	D128 2345	Steps
10th	D109 2345	Steps	20th	D119 2345	Steps	30th	D129 2345	Steps

Use the following settings for each component.

Title
Figures tab ⇒Text
String: April step count history
Text size: 35
Figure position and size
(Displayed at the lower left after clicking on the shape)
X:200 Y:10 Width:245

Date
Figures tab ⇒Text
String:01 day
Text size: 20
Figure position and size (reference setting)
(Displayed in the lower left corner after clicking on a part)
X:45 Y:70 Width:42

Steps display (D100)
Objects" tab
⇒Numeric Display/Input ⇒Numerical Display
Device: D100
Numerical size: 20
Integer part digits: 5 digits
Preview value: 12345
Figure: Rectangle_Solid_Border_Width_Fixed: Rect_7
Part location and size (reference setting)
(Displayed in the lower left corner after clicking on a part)
X:90 Y:60 W:80 H:36

Unit

Figures tab ⇒Text

String: step

Text size: 20

Figure position and size (reference setting)

(Displayed in the lower left corner after clicking on a part)

X:175 Y:70 Width:20

Continuous copy setting (01 day)

X direction:1

Y direction: 10,

Y-directional spacing: 5

Increment setting: batch, 1

Continuous copy setting (01 day to 10 days)

X direction:3

Y direction: 1,

X-directional spacing: 50

Increment setting: batch, 10

Exercise 6 Solution

1. After double-clicking on "New", enter screen number: 16, title: Question 6, and press "OK" to create a new screen.

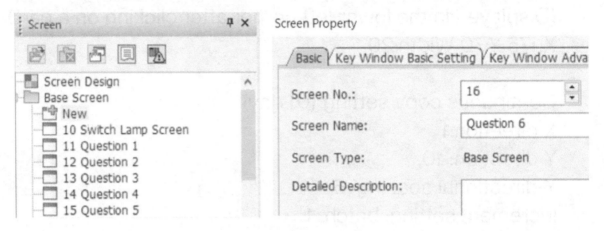

2. create a title and a display of the number of steps taken on 01 day (D100) using the text of the shape and the numerical display of the object.

Title
"Figure" tab => Text
String: April step count history   Text size: 35

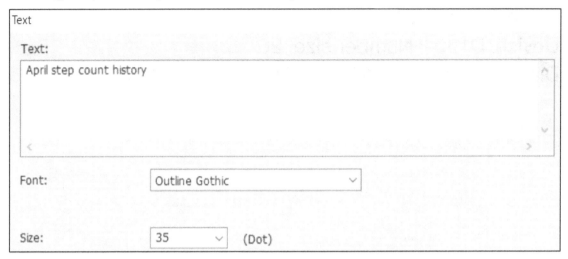

Figure position and size (reference setting)
(Displayed at the lower left after clicking on the shape)
X:200 Y:10 Width:245

Date
Figure" tab => Text
String:01 day   Text size: 20

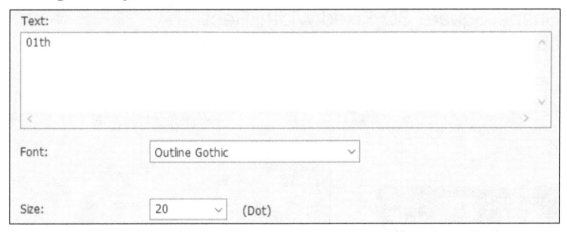

Figure position and size (reference setting)
(Displayed in the lower left corner after clicking on a part)
X:45 Y:70 Width:42

Steps display (D100)
"Object" tab
⇒Numeric Display/Input ⇒Numeric Display
Basic Settings (Devices)
Device: D100　Number size: 20
Digits(Integral): 5 digits　Preview value: 12345

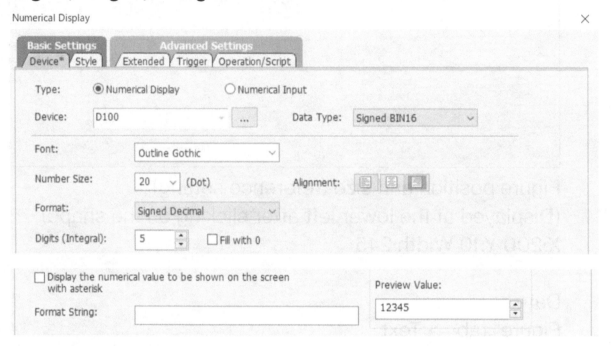

Basic Settings (Style)
Shape: Square_3D_Fixed Width: Rect_7

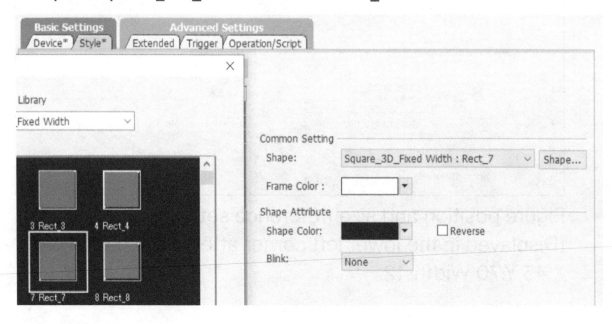

176

Part location and size (reference setting)
(Displayed in the lower left corner after clicking on a part)
X:90 Y:60 W:80 H:36

Unit
"Figure" tab => Text
String: Steps   Text size: 20

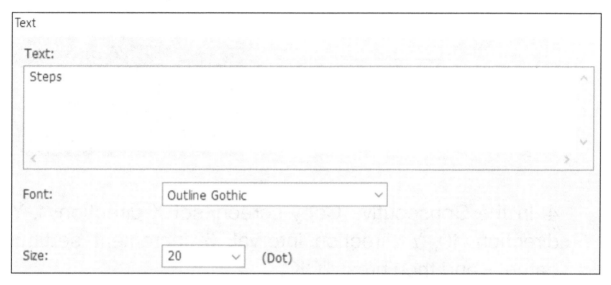

Figure position and size (reference setting)
(Displayed in the lower left corner after clicking on a part)
X:175 Y:70 Width:20

3. Select all parts except the title, then select "Edit" tab ⇒ Consecutive Copy, or right-click on the part ⇒ Consecutive Copy.

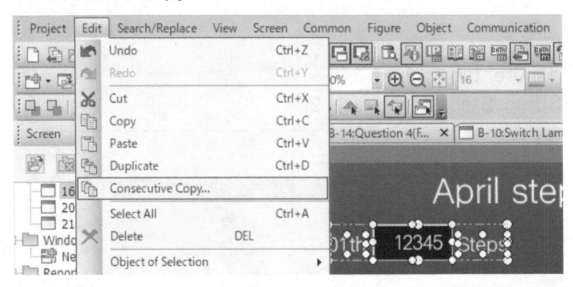

4. In the Consecutive Copy screen, set X direction: 1, Y direction: 10, Y direction interval: 5, increment setting: batch, 1, and then press "OK".

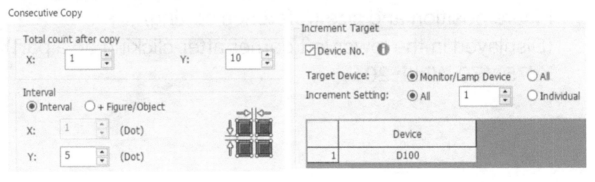

5. After Consecutive copying, check the device display and confirm that the devices are D100 to D109 from the top to the bottom.

6. Double-click on the date string and change it as shown in the following screen.

7. Select all parts from 01 to 10 and again select Consecutive copy.

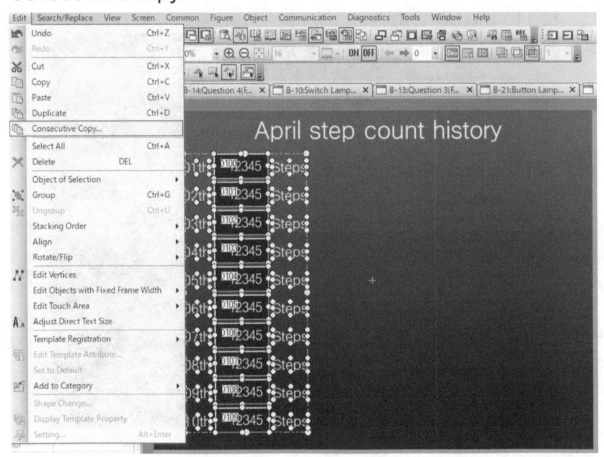

8. In the Consecutive Copy screen, set X direction: 3, Y direction: 1, X direction interval: 15, increment setting: All, 10, and then press "OK".

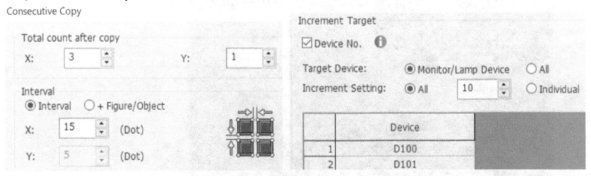

9. After Consecutive copying, check the device display and confirm that the second row from the left is D110 to D119 from the top, and the third row from the left is D120 to D129 from the top.

April step count history

01th	D100 2345	Steps	01th	D110 2345	Steps	01th	D120 2345	Steps
02th	D101 2345	Steps	02th	D111 2345	Steps	02th	D121 2345	Steps
03th	D102 2345	Steps	03th	D112 2345	Steps	03th	D122 2345	Steps
04th	D103 2345	Steps	04th	D113 2345	Steps	04th	D123 2345	Steps
05th	D104 2345	Steps	05th	D114 2345	Steps	05th	D124 2345	Steps
06th	D105 2345	Steps	06th	D115 2345	Steps	06th	D125 2345	Steps
07th	D106 2345	Steps	07th	D116 2345	Steps	07th	D126 2345	Steps
08th	D107 2345	Steps	08th	D117 2345	Steps	08th	D127 2345	Steps
09th	D108 2345	Steps	09th	D118 2345	Steps	09th	D128 2345	Steps
10th	D109 2345	Steps	10th	D119 2345	Steps	10th	D129 2345	Steps

10. batch change the string: select the 01 to 09 days in the second column, then select the "Search/Replace" tab ⇒Data Browser.

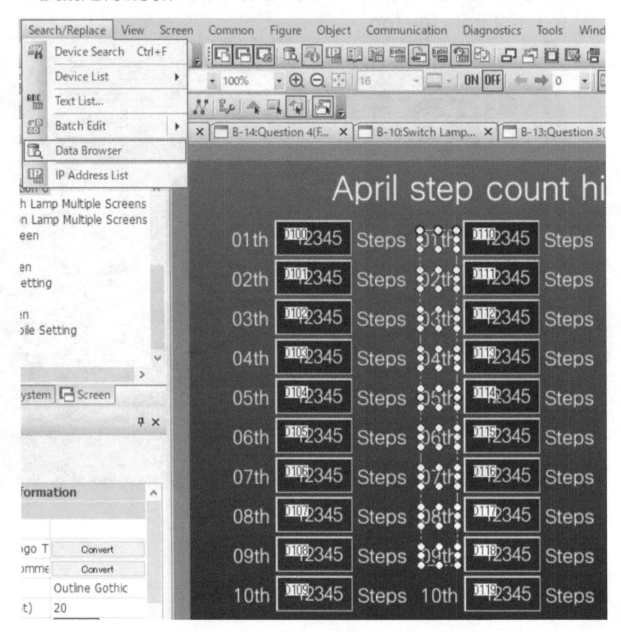

11. Select "Option" ⇒ "Text Batch Edit" in the Data Browser, enter "0" for the string to be changed and "1"for the string after the change, and then press "Replace".

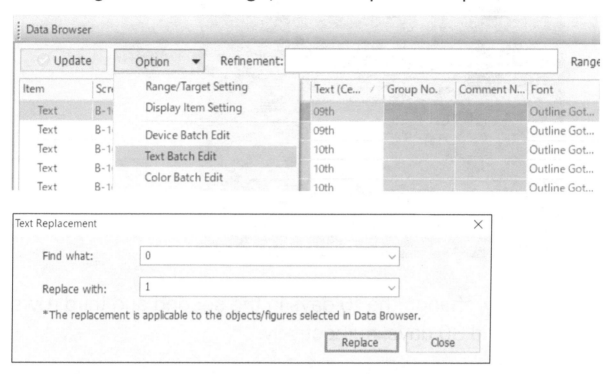

12. Confirm that column 2, days 01 through 09 have been changed to days 11 through 19.

13. Using the same procedure, change column 3, days 01 through 09, to days 21 through 29.

April step count history

01th	D1002345	Steps	11th	D1102345	Steps	21th	D1202345	Steps
02th	D1012345	Steps	12th	D1112345	Steps	22th	D1212345	Steps
03th	D1022345	Steps	13th	D1122345	Steps	23th	D1222345	Steps
04th	D1032345	Steps	14th	D1132345	Steps	24th	D1232345	Steps
05th	D1042345	Steps	15th	D1142345	Steps	25th	D1242345	Steps
06th	D1052345	Steps	16th	D1152345	Steps	26th	D1252345	Steps
07th	D1062345	Steps	17th	D1162345	Steps	27th	D1262345	Steps
08th	D1072345	Steps	18th	D1172345	Steps	28th	D1272345	Steps
09th	D1082345	Steps	19th	D1182345	Steps	29th	D1282345	Steps
10th	D1092345	Steps	10th	D1192345	Steps	10th	D1292345	Steps

14. lastly, change the 10 days in the second and third rows to 20 and 30 days, respectively.

April step count history

01th	D1002345	Steps	11th	D1102345	Steps	21th	D1202345	Steps
02th	D1012345	Steps	12th	D1112345	Steps	22th	D1212345	Steps
03th	D1022345	Steps	13th	D1122345	Steps	23th	D1222345	Steps
04th	D1032345	Steps	14th	D1132345	Steps	24th	D1232345	Steps
05th	D1042345	Steps	15th	D1142345	Steps	25th	D1242345	Steps
06th	D1052345	Steps	16th	D1152345	Steps	26th	D1252345	Steps
07th	D1062345	Steps	17th	D1162345	Steps	27th	D1262345	Steps
08th	D1072345	Steps	18th	D1172345	Steps	28th	D1272345	Steps
09th	D1082345	Steps	19th	D1182345	Steps	29th	D1282345	Steps
10th	D1092345	Steps	20th	D1192345	Steps	30th	D1292345	Steps

Chapter6: Frequently Used Objects

6-1 Switch

There are several types of object switches. The most used of these switches are switches, bit switches, and screen switching switches.

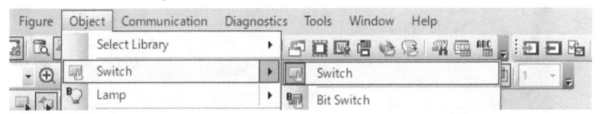

The bit switches and screen switching switches have been described in previous chapters, so this section describes the switches.

After selecting the "Object" tab ⇒Switch ⇒Switch, click on the screen to place the part. Then double-click on the part to display the switch settings screen.

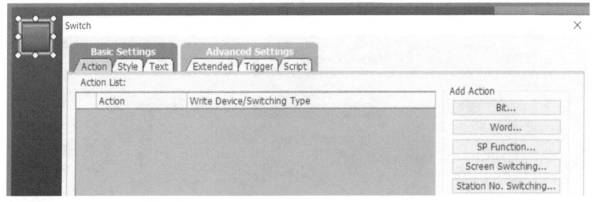

This switch component allows multiple items to be added to the operation list in the Add Operation column on the right. (One switch can have multiple actions.)

For example, suppose you have a fan with buttons off, 1 : low (X30), 2 : medium (X31), and 3 : strong (X32).

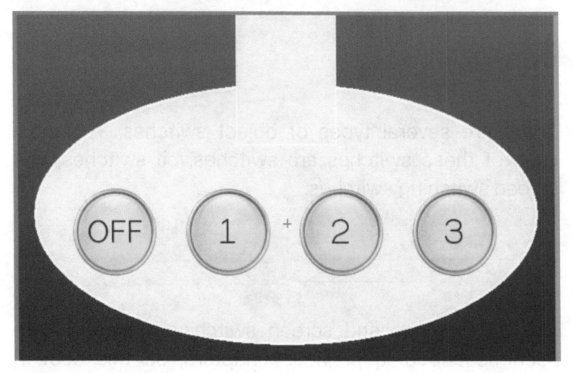

If you want to turn off all low (X30), medium (X31), and strong (X32) when you press the fan's off button, first select the additional bit of operation and press "OK" as Device: X0030, Action: bit reset.

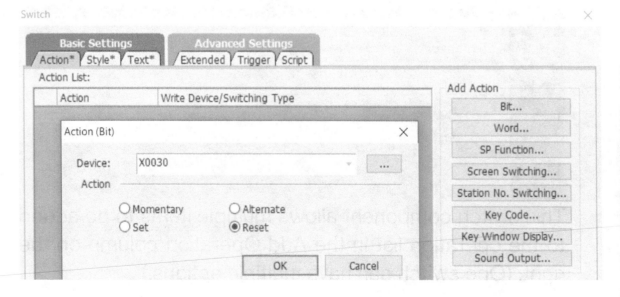

This will add the bit reset for X0030 to the operation list.

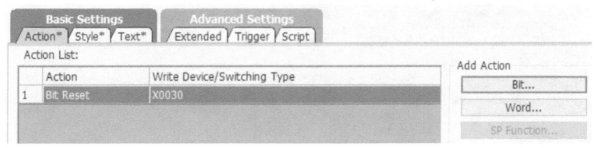

Follow the same procedure to add bit resets for X0031 and X0032 and press "OK". This completes the setting of the cut button.

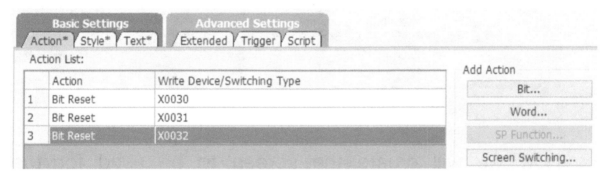

Next, in the case of the fan's low (X30), medium (X31), and strong (X32) buttons, for example, pressing the low button will turn on X30 and turn off X31 and X32. This operation is also possible by setting multiple operations for the switch.

Weak Button Setting

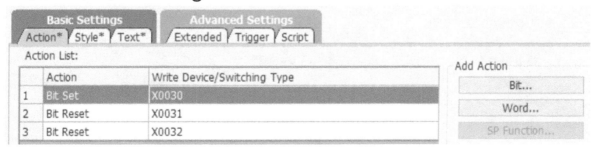

Middle Button Setting

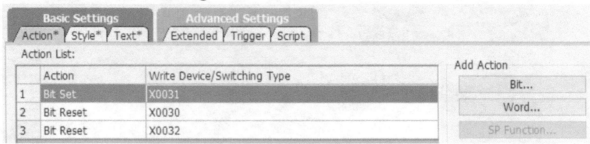

Strong button setting

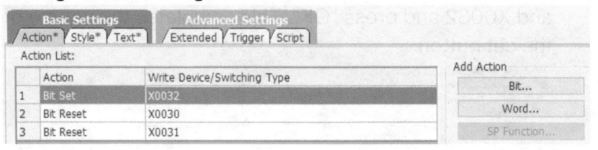

Now we will create the screen to be used for the simulation.

1. Double-click "New" on the base screen, enter screen number: 22, title: Fan button, and press "OK" to create a new screen.

Screen Property

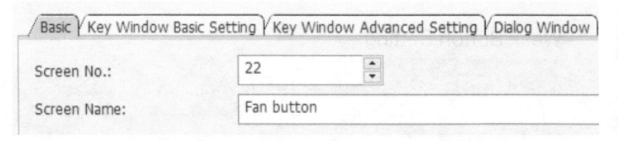

2. First, go ahead and create thenon-button parts of the fan.

"Figure" ⇒Circle
Line width: 3Dot
Pattern: 37
Background color: light gray (one shade below white)

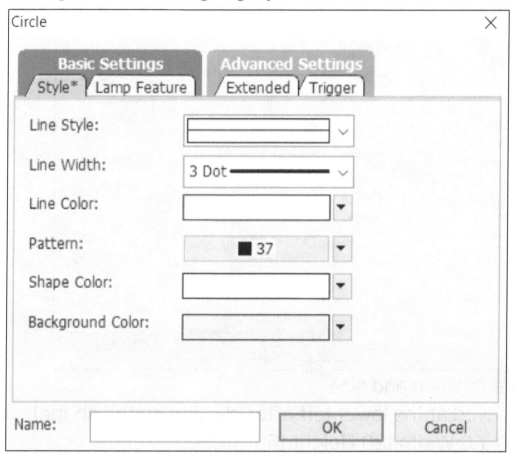

Figure position and size
(Displayed at the lower left after clicking on the shape)
X:20 Y:80 Width:600 Height:300

Figure" ⇒Rectangle
Pattern: 37
Background color: light gray (one shade below white)

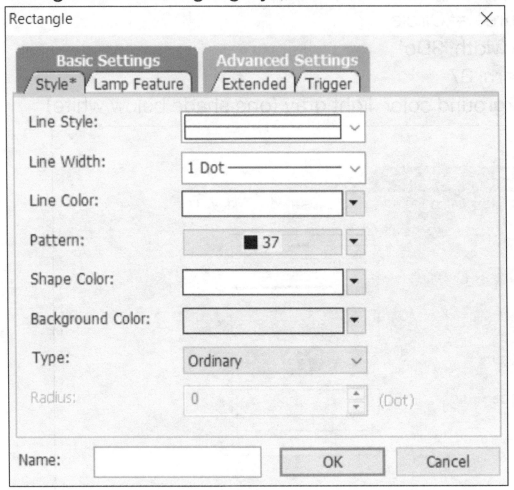

Figure position and size
(Displayed at the lower left after clicking on the shape)
X:260 Y:0 Width:120 Height:150

3. Create each button of the fan.

button to turn off.
Object: Switch
Basic Settings (Action)
Action List
Bit reset:X0030.
Bit reset:X0031.
Bit reset:X0032.
Lamp function: key touch status

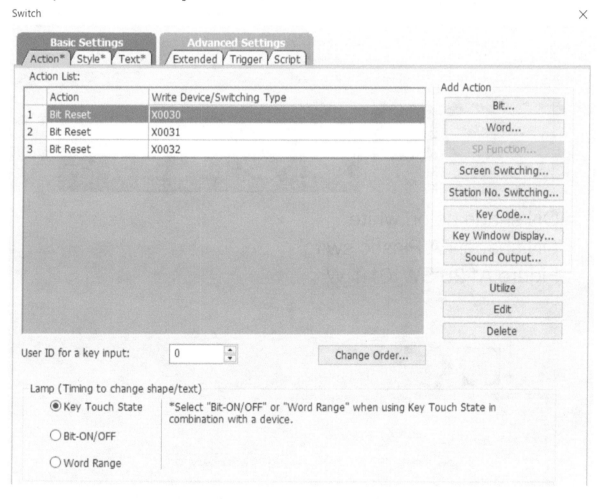

Basic Settings (Style)
OFF shape color: white
OFF shape: FA plastic switch
Figure A1 25 SW_01_0_W

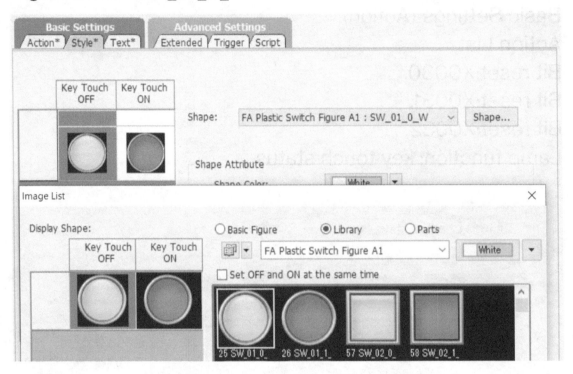

ON shape color: white
ON shape: FA Plastic switch
Figure A1 26 SW_01_1_W

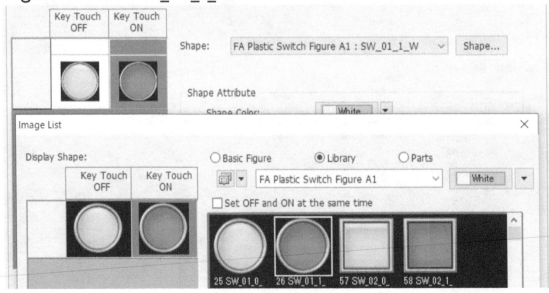

[Basic settings (text)]

Text type OFF=ON check: OFF

Text size: 36

OFF Text: OFF

OFF Text color: black

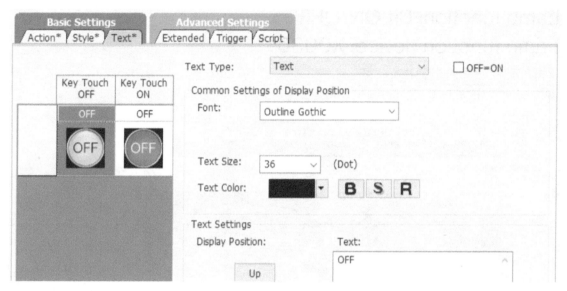

Text size: 36

ON Text: OFF

ON Text color: white

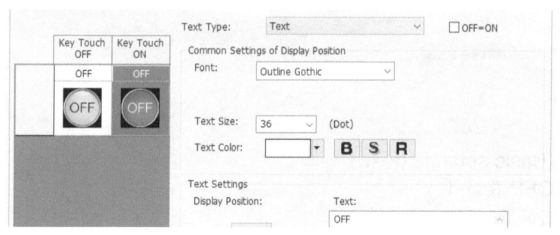

Part location and size

(Displayed in the lower left corner after clicking on a part)

X:65 Y:200 Width:100 Height:100

Weak button
Copy ⇒Paste
Basic Settings (Action)
Action List
Bit set:X0030.   Bit reset:X0031.   Bit reset:X0032.
Lamp function: Bit ON/OFF
Lamp function device: X0030

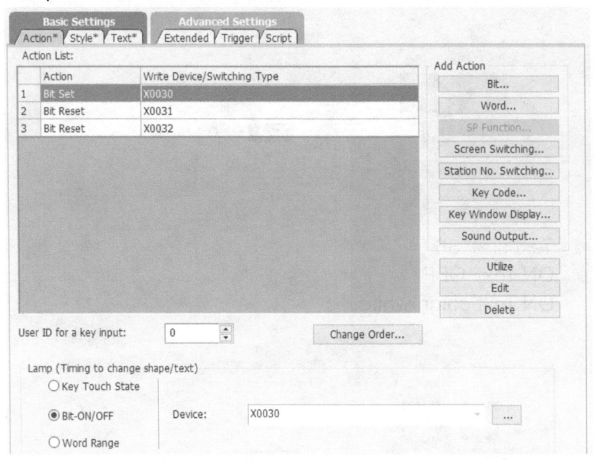

[Basic settings (text)]
OFF Text: 1
ON Text: 1
Part location and size
(Displayed in the lower left corner after clicking on a part)
X:200 Y:200 Width:100 Height:100

Middle button

Copy ⇒Paste

Basic Settings (Action)

Action List

Bit Set:X0031.　Bit reset:X0030.　Bit reset:X0032.

Lamp function: Bit ON/OFF

Lamp function device: X0031

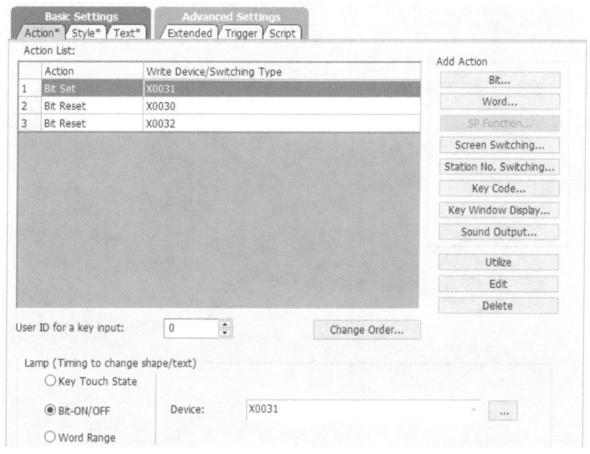

[Basic settings (text)]

OFF Text: Medium

ON Text: Medium

Part location and size

(Displayed in the lower left corner after clicking on a part)

X:335 Y:200 Width:100 Height:100

Strong button
Copy ⇒Paste
Basic Settings (Action)
Action List
Bit Set:X0032.　Bit reset:X0030.　Bit reset:X0031.
Lamp function: Bit ON/OFF
Lamp function device: X0032

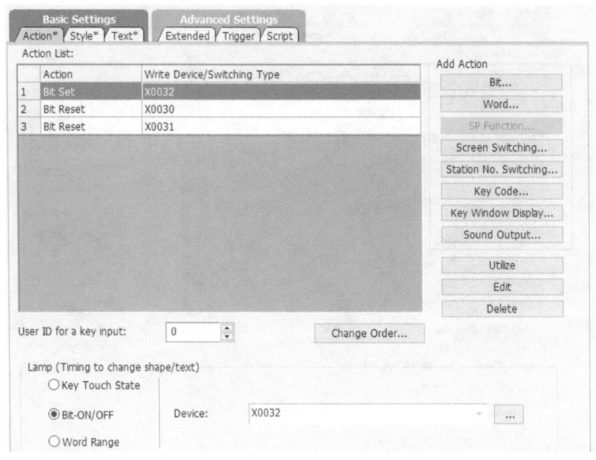

[Basic settings (text)]
OFF Text: Strong
ON Text: Strong
Part location and size
(Displayed in the lower left corner after clicking on a part)
X:470 Y:200 Width:100 Height:100

Also, add a "Next" button on Screen 14 to go to Screen 22 and a "Previous" button on Screen 22 to go to Screen 14, respectively, so that the user can move to Screen 22 during the simulation.

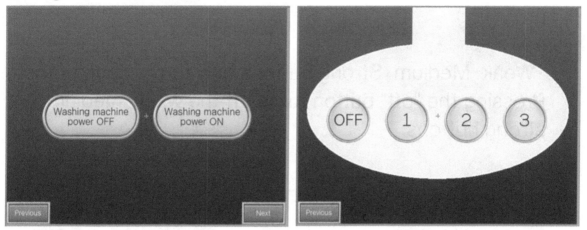

When you have finished creating the screen, start the PLC program used in the past simulations, start the simulator for the screen, and confirm the following operations. (It is not necessary to create the PLC program.)

·Weak·Medium·Strong buttons only turn on one of them. Pressing the "off" button turns off the weak, medium, and strong buttons.

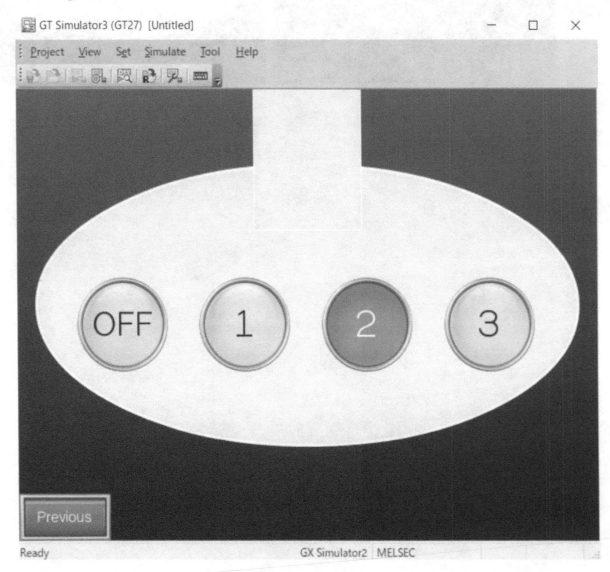

6-2 Word lamp

The next section describes word lamps.
While bit lamps could only represent two colors, ON and OFF, word lamps can represent three or more colors.

For example, in standby (D200=0), the lamp is blue,

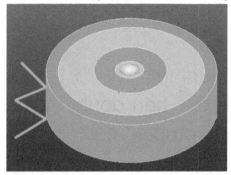

During operation (D200=1), the lamp is green,

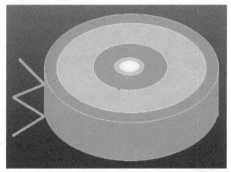

Lamp is red when an abnormality is occurring (D200=99)

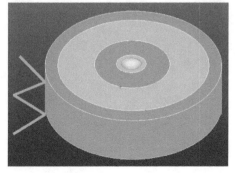

Suppose there is a robot vacuum cleaner that becomes.
Create this lamp using a word lamp.

First, after double-clicking New on the base screen, enter screen number: 23, title: Robot Vacuum Cleaner Lamp, and press "OK" to create a new screen.

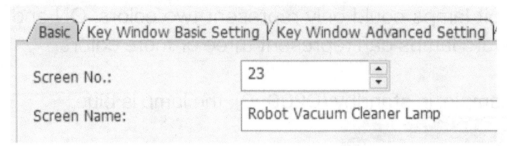

After selecting the "Object" tab ⇒Lamp ⇒Word Lamp, click on the screen to place the part. Then double-click on the part to display the word lamp setup screen.

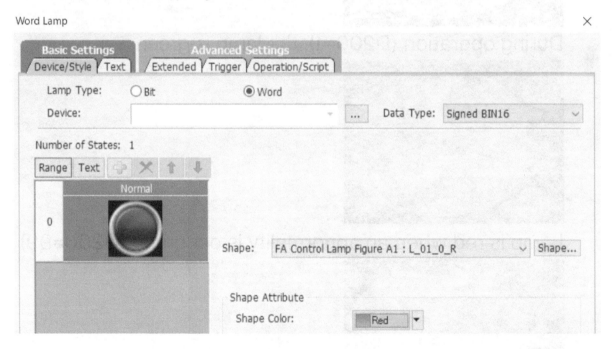

First, set the lamp to be blue during standby (D200=0). First, enter D200 in Device. Then click on "Shape..." in Shape (P), select "Blue" from the list of colors, then select "FA Control Lamp Figure A1" from the list selection in Library, click on "2 L_01_1_B" and press "OK" in the lower right corner.

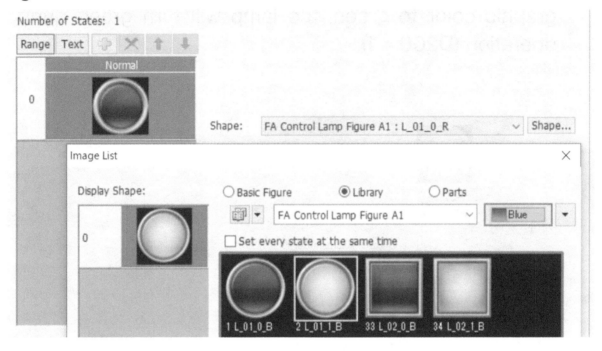

This will now be a blue lamp when in standby (D200=0).

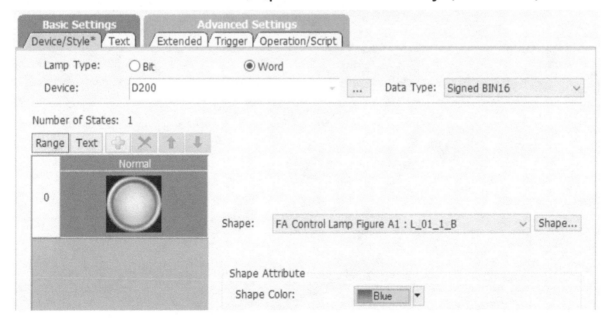

Next, set the lamp to be green during operation (D200=1). Pressing the plus sign under the number of conditions: 1 on the setup screen will set the number of conditions to 2 and add a lamp that lights up under the condition $V == 1. This $V represents the value of device D200, so it will light up when D200 reaches 1. Now, by changing the graphic color to green, the lamp will turn green during operation (D200 = 1).

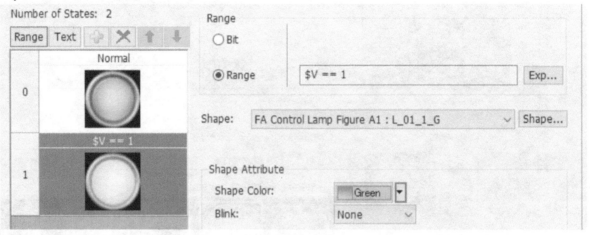

Finally, set the lamp to turn red during an error (D200=99). Press the plus sign under the number of conditions: 2 on the settings screen to set the number of conditions to 3 and add a lamp that lights up under the condition $V == 2. This time, we want to turn red when D200 = 99, so we will change $V == 2 written in the range. Select "Range..." to display the range input screen. Change the B constant to 99 and press "OK.

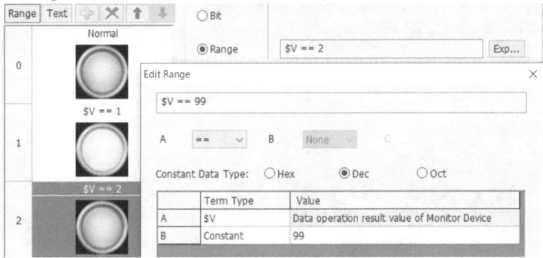

Then, by changing the graphic color to a reddish color, the lamp will turn red when an abnormality is occurring (D200=99). This completes all settings.

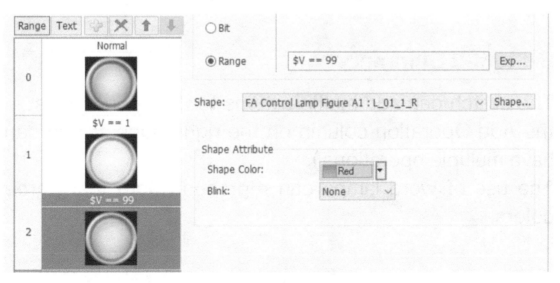

You can check the operation on the screen by changing the condition No. under the "Tools" tab.

6-1 to 6-2 Summary

The switch can add multiple items to the operation list in the Add Operation column on the right. (One switch can have multiple operations.)
The use of word lamps can represent more than three colors.

Exercise 7 (Washing machine standard course time display)

Create a new screen number 30, question 7, and use the switches to create a screen where pressing the normal course button on the washing machine will put wash (D210): 20 minutes, rinse (D220): 15 minutes, dry(D230): 10 minutes, and pressing the reset button will set all setting times to 0 minutes. (Please also add a button to move between screens 22-30)

normal course button OFF

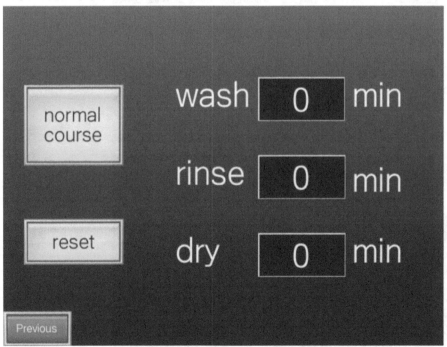

normal course button ON

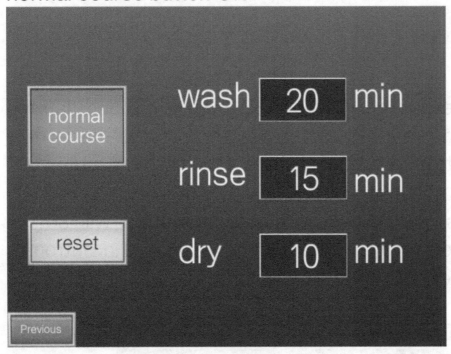

Reset button ON

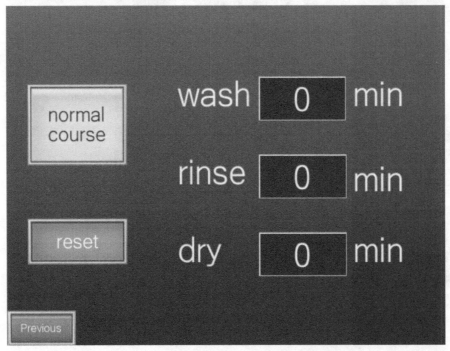

Exercise 7 Solution

normal course button
Object tab ⇒Switch ⇒ Switch

Basic Settings (Action)
Click on "Word" to add an action,
Device: D210 Constant: 20
Click on "Word" to add an action,
Device: D220 Constant: 15
Click on "Word" to add an action,
Device: D230 Constant: 10

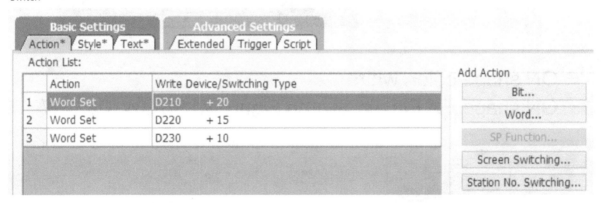

After that, you are free to set up as you wish.

(Reference setting)
Basic Settings (Style)
OFF shape color: white
OFF shape: FA Plastic Switch Figure A1 57 SW_02_0_W

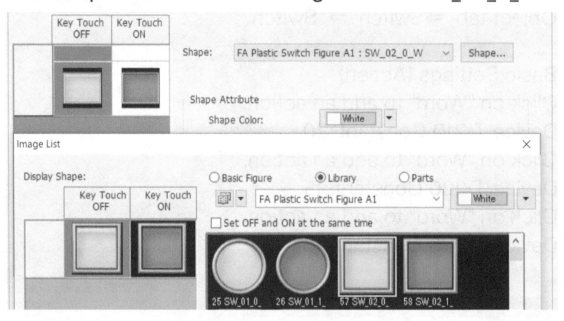

ON shape color: white
ON shape: FA Plastic Switch Figure A1 58 SW_02_1_W

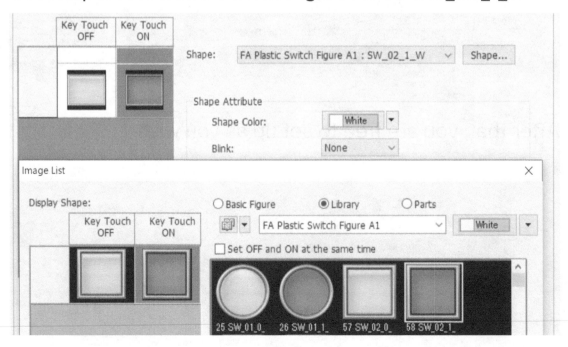

[Basic settings (text)]
Text type OFF=ON check: OFF
Text size: 28
OFF text: normal course   Text color: black

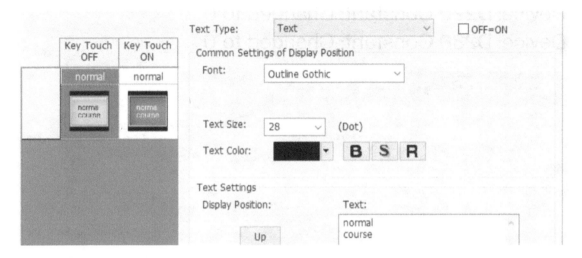

ON text: normal course   Text color: White

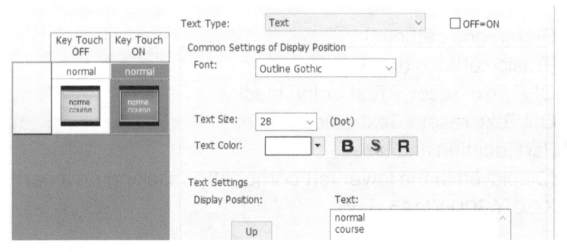

Part location and size
(Displayed in the lower left corner after clicking on a part)
X:30 Y:110 W:145 H:115

Reset button.

Copy ⇒Paste normal course Button

Basic Settings (Action)

Device: D210 Changed to constant: 0

Device: D220 Constant: Changed to 0

Device: D230 Constant: Changed to 0

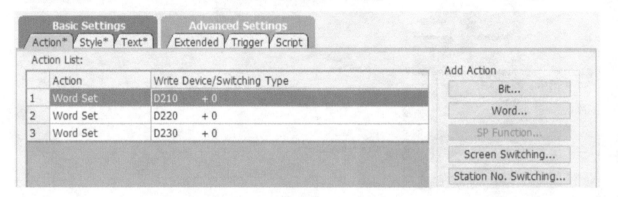

After that, you are free to make any settings you wish.

(Reference settings).

[Basic settings (text)]

OFF Text: reset   Text color: black.

ON Text: reset   Text color: white.

Part location and size

(Displayed in the lower left corner after clicking on a part)

X:30 Y:300 W:145 H:65

Each Text of wash/rinse/dry/minute.
"Figure" tab => Text
Text size: 48

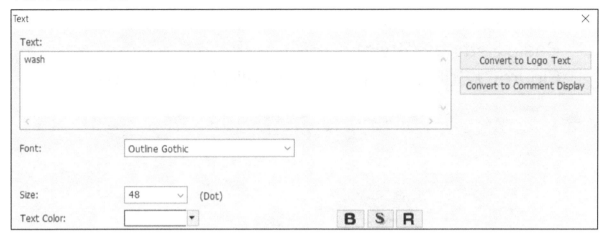

wash display area
"Object" tab => Numeric Display/Input
⇒Numerical Display
Basic Settings (Devices)
Device: D210

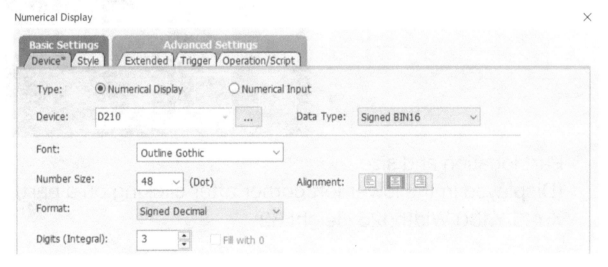

*After that, you are free to set up as you wish.

(Reference setting)
Basic Settings (Devices)
Numerical size: 48
Integer part digits:3
Basic Settings (Style)
Figure:71 Square_Solid_Border_Width_Fixed 7 Rect_7

Part location and size
(Displayed in the lower left corner after clicking on a part)
X:415 Y:100 Width:125 Height:62

rinse /dry display area
Copy ⇒Paste
rinse is device: D220.
dry is device: D230.

Other
Added "Next" button on screen 22.
Added "Previous" button on screen 30.

When you have finished creating the screen, start the PLC program used in the past simulations, start the simulator for the screen, and confirm the following operations. (It is not necessary to create the PLC program.)

Press the standard course button on thewashing machine, wash: 20 min. rinse: 15 min. dry: 10 minutes enter.

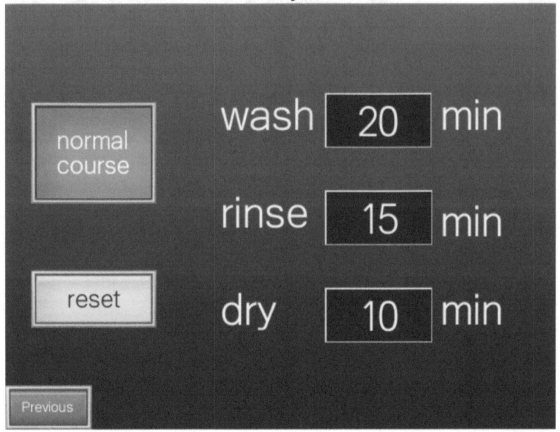

Press the reset button on the washing machine,
wash: 0 min. rinse: 0 min. dry: 0 minutes enter.

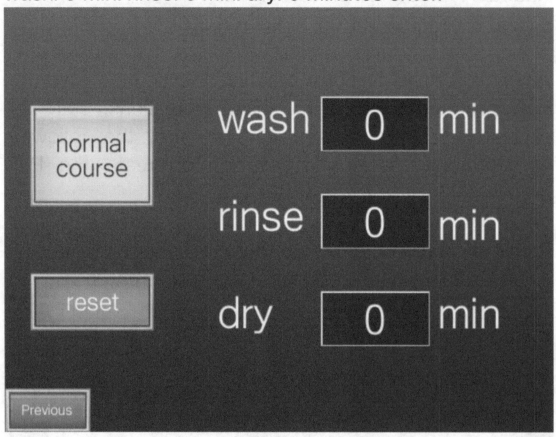

Exercise 8 (Lighting color change button)

Create a new screen number 31, question 8, and use the word lamp to create a screen thatrepeats orange (D240=0) ⇒purple (D240=1) ⇒pink (D240=2) ⇒cyan (D240=3) ⇒green (D240=4) when the Change Lighting Color button is pressed. (Please also add a button to move between screens 30-31)

Initial state (D240=0)

↓ Lighting color change button ON (D240=1)

↓ Lighting color change button ON (D240=2)

↓ Lighting color change button ON (D240=3)

↓ Lighting color change button ON (D240=4)

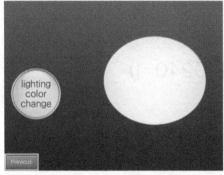

↓ Lighting color change button ON (D240=0)

Exercise 8 Solution

Lighting
"Object" tab => Lamp ⇒ Word Lamp

Basic Settings (Device/Style)
Device: D240

<Condition 0>
Shape color: Orange
Shape: FA Control_N Lamp figure A1 10 L_01_1_O

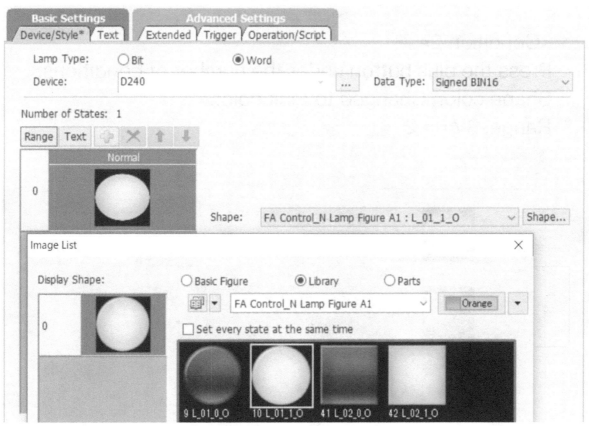

<Condition 1>

Press the plus button under the number of conditions.

Shape color: Changed to purple.

Range: $V == 1

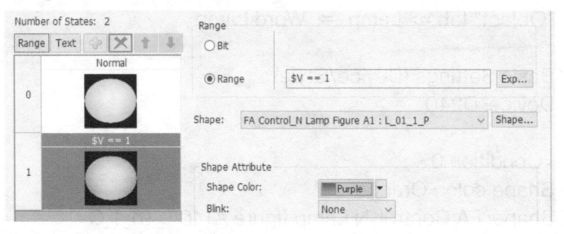

<Condition 2>

Press the plus button under the number of conditions.

Shape color: Changed to Pink color.

Range: $V == 2

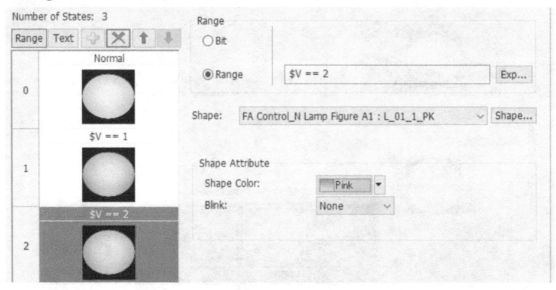

<Condition 3>

Press the plus button under the number of conditions.

Shape color: Changed to Cyan.

Range: $V == 3

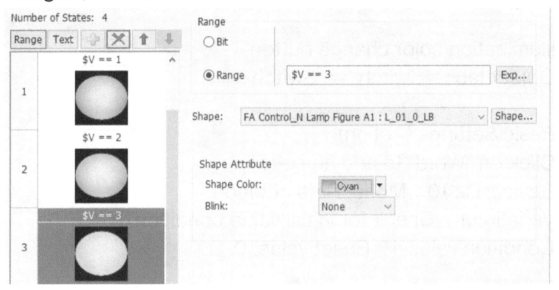

<Condition 4>

Press the plus button under the number of conditions.

Shape color: Changed to green.

Range: $V == 4

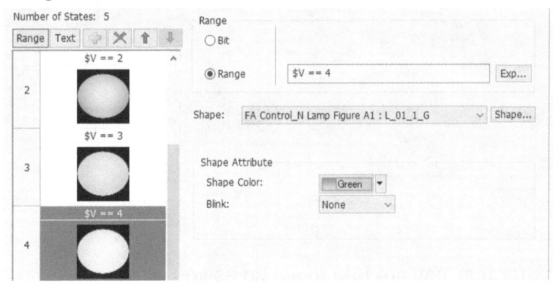

(Reference setting)
Part location and size
(Displayed in the lower left corner after clicking on a part)
X:290 Y:100 W:300 H:250

Illumination color change button
Object tab ⇒Switch ⇒ Switch

Basic Settings (Action)
Click on "Word" to add an action,
Device: D240   Mode: Data addition
Variation:1.   Check for initial value condition.
Condition value:4   Reset value: 0.

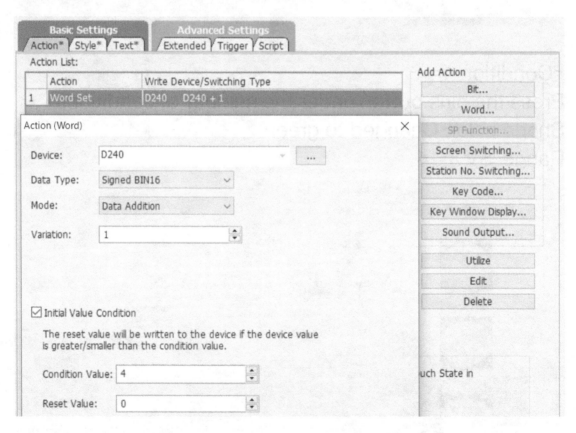

*After that, you are free to set up as you wish.
(Reference setting)

Basic Settings (Style)
OFF shape color: white
OFF shape: FA plastic switch Figure A1 25 SW_01_0_W

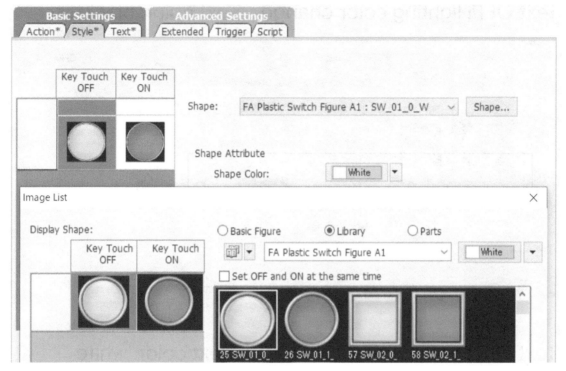

ON shape color: white
ON shape: FA Plastic switch Figure A1 26 SW_01_1_W

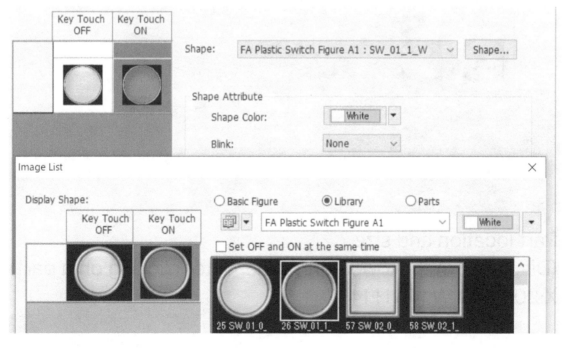

[Basic settings (text)]
Text type OFF=ON check: OFF
Text size: 28
Text OFF: lighting color change   Text color: black

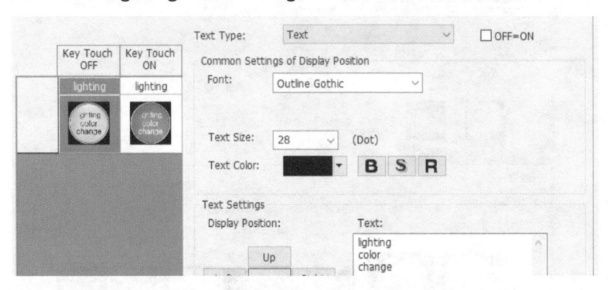

Text ON: lighting color change   TRext color: white

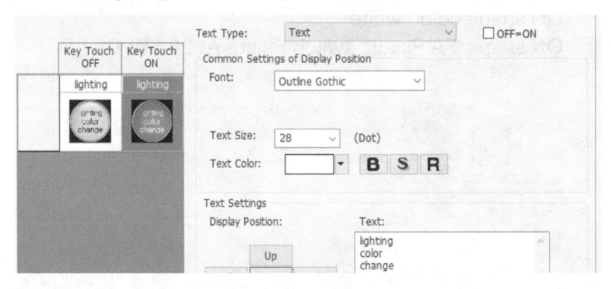

Part location and size
(Displayed in the lower left corner after clicking on a part)
X:20 Y:190 W:145 H:145

Other
Added "Next" button on screen 30.
Added "Previous" button on screen 31.

When you have finished creating the screen, start the PLC program used in the past simulations, start the simulator for the screen, and confirm the following operations. (It is not necessary to create the PLC program.)

When the "Change Lighting Color" button is pressed, the following sequence is repeated: orange (D240=0) ⇒ purple (D240=1) ⇒pink (D240=2) ⇒cyan (D240=3) ⇒ green (D240=4) ⇒orange (D240=0)...

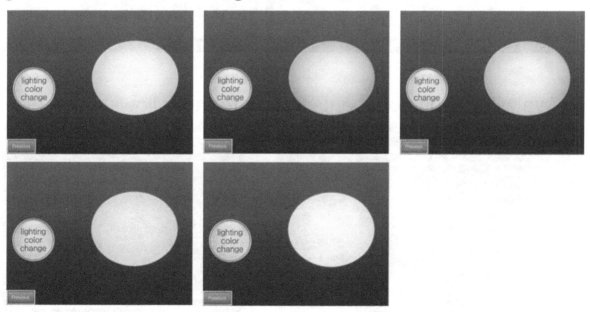

6-3 Bit comments

Bit comments are used to switch comments on and off, rather than switching the color of the lamp like a bit lamp. For example, if you want the microwave monitor to alternately display 800W and 1000W each time you press the 800W/1000W switch button on the microwave oven, use bit comments.

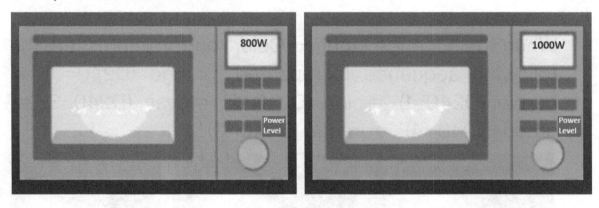

We will use bit comments to create the following screen.

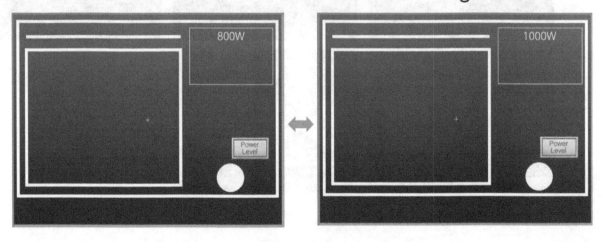

1. First, create a new screen with screen number: 32 and title: Microwave Oven.

Screen Property

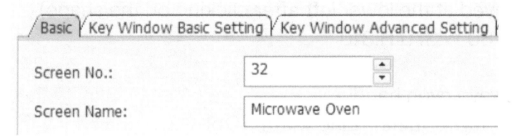

2. create figures other than the bit lamp and W number switch.

Microwave Oven Outer Frame

Figure tab => Rectangle   Line width: 7Dot

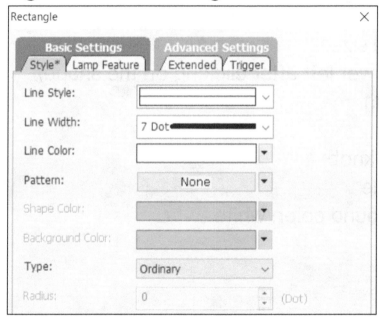

Figure position and size

(Displayed at the lower left after clicking on the shape)

X:10 Y:15 W:620 H:400

Microwave heating point

Copy ⇒Paste the outer microwave frame

Figure position and size

(Displayed at the lower left after clicking on the shape)

X:30 Y:80 W:370 H:310

Microwave oven handle

Figure" tab => Line   Line width: 7Dot

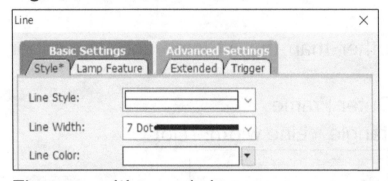

Figure position and size

(Displayed in the lower left after clicking on the shape)

X:30 Y:50 W:370 H:1

Microwave control knob

Figure" tab => Circle

Pattern: 37 Background color: white

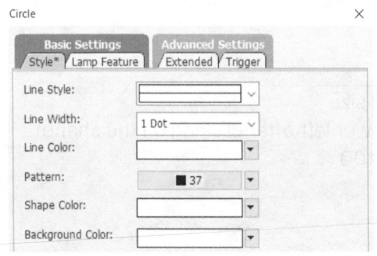

Figure position and size
(Displayed in the lower left corner after clicking on the shape)
X:485 Y:340 Width:65 Height:65

Microwave monitor frame
Figure tab => Rectangle
Figure position and size
(Displayed at the lower left after clicking on the shape)
X:420 Y:30 Width:200 Height:130

3. Create a bit lamp. After selecting the "Object" tab ⇒ Comment Display ⇒Bit Comment, click on the screen to place the part.

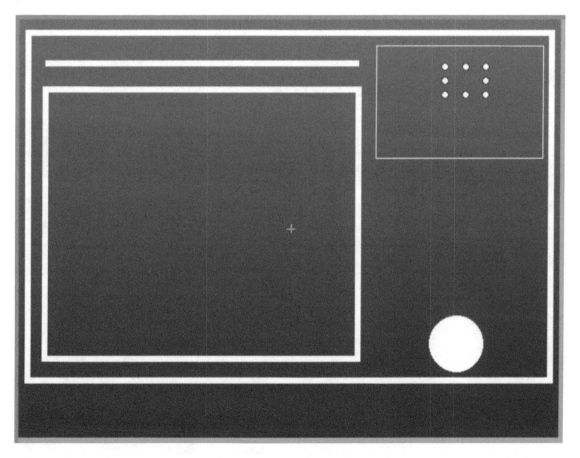

4. Double-click on the part to display the bit comment setting screen and make the following settings.

Basic Settings (Device/Style)
Device: X0040

Bit Comment Display

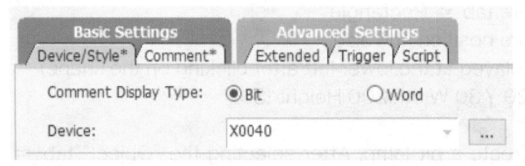

Basic Settings (Comment)
OFF=ON check: OFF   Display Type: Text
OFF:800W   Text size: 26

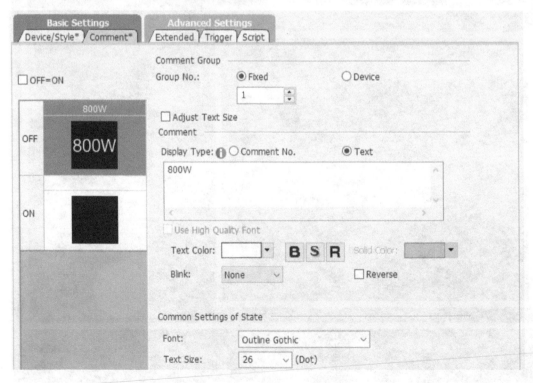

ON: 1000W   Text size: 26

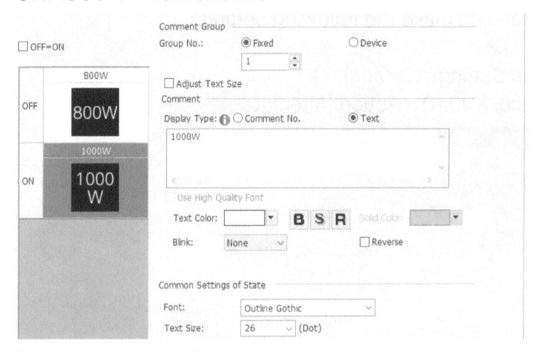

Part location and size
(Displayed in the lower left corner after clicking on a part)
X:470 Y:30 Width:100 Height:40

5. Create a power level switch. After selecting "Object" tab ⇒Switch ⇒Bit Switch, click on the screen to place the parts.

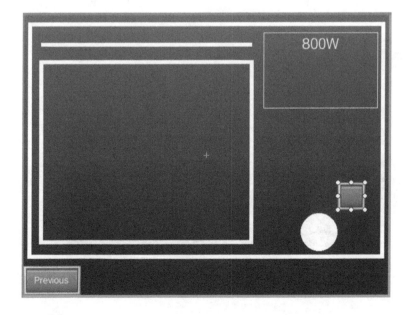

6. Double-click the part to display the bitswitch setting screen and make the following settings.

Basic Settings (Device)
Device: X0040   Action: Alternate

Basic Settings (Style)
OFF shape color: white
OFF shape: FA control switch Figure A1 57 SW_02_0_W

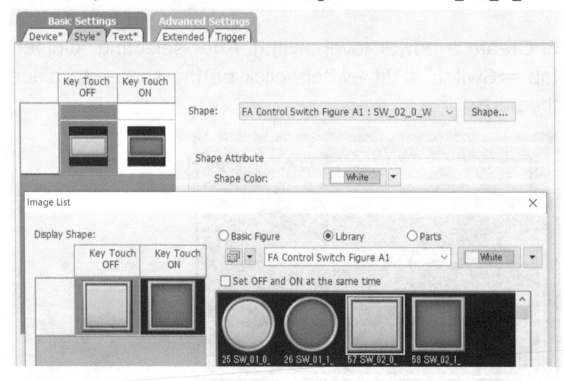

ON shape color: white
ON shape: FA control switch Figure A1 58 SW_02_1_W

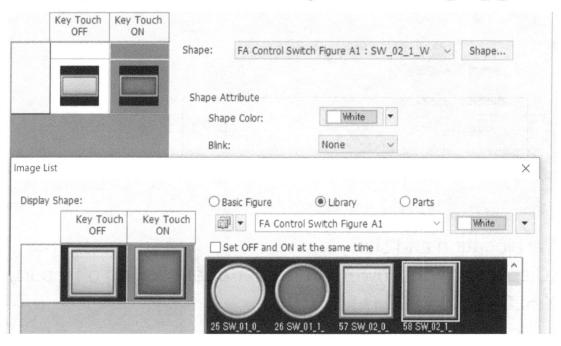

[Basic settings (text)]
Text type OFF=ON check: OFF
Text size: 16
Text OFF: Power Level.   Text color: black

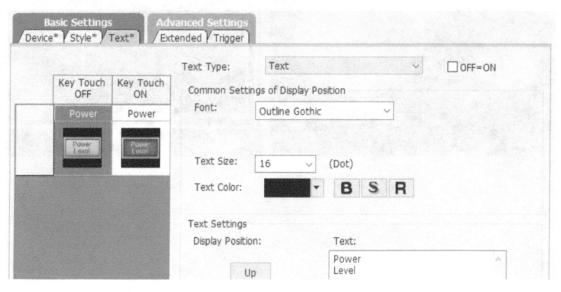

Text ON: Power Level.　Text color: white

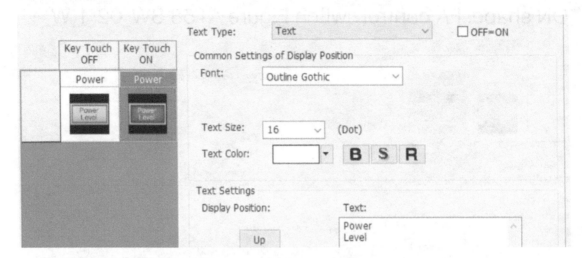

Part location and size
(Displayed in the lower left corner after clicking on a part)
X:520 Y:280 W:90 H:50

7. add a "Next" button on Screen 31 and a "Previous" button on Screen 32.

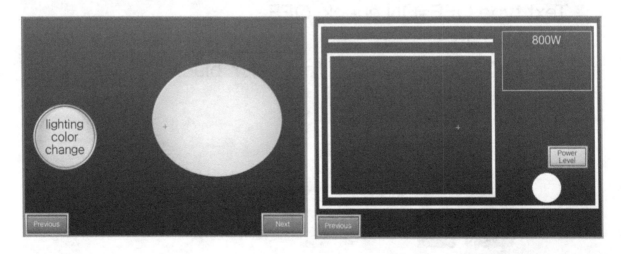

8. After creating the screen, start the PLC program used in the past simulations, start the simulator for the screen, and confirm the following operations. (It is not necessary to create the PLC program.)

Each time the "Power Level" switch is pressed, itswitches between 800W and 1000W.

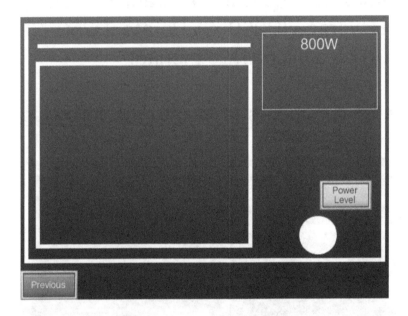

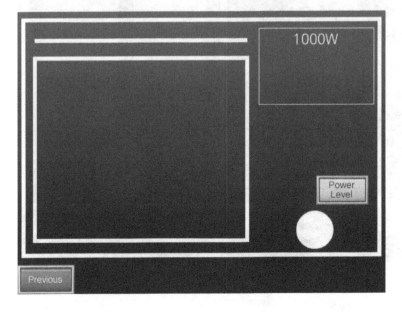

6-4 Word comments

While bit comments switch comments by ON/OFF, word comments can switch multiple comments by the number in the specified device.

For example, use word comment if you want to display Popcorn (D250=1), Beverage (D250=2), Potato (D250=3), and Reheat (D250=4) on a microwave monitor.

We will be using word comments to create the following screens.

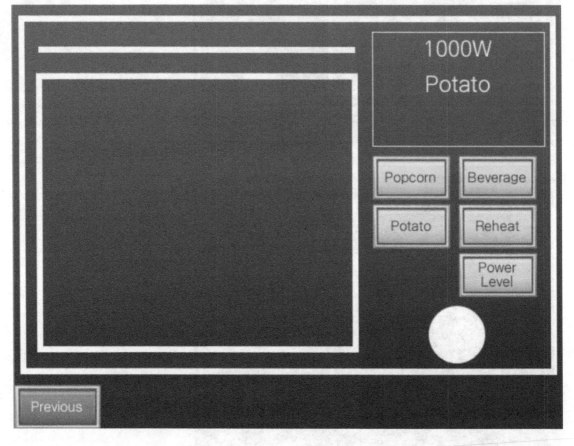

1. create a word lamp. After selecting the "Object" tab ⇒ Comment Display ⇒Word Comment, click on the screen to place the component.

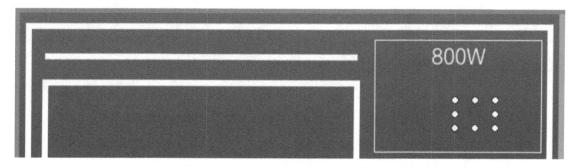

2. Double-click on the part, and the word comment setting screen will appear. After entering Device: D250 in the Basic Settings (Device/Style), press the Plus button under Number of Conditions: 1 four times to set the number of conditions to 5, and conditions D250=1 to D250=4 will be added.

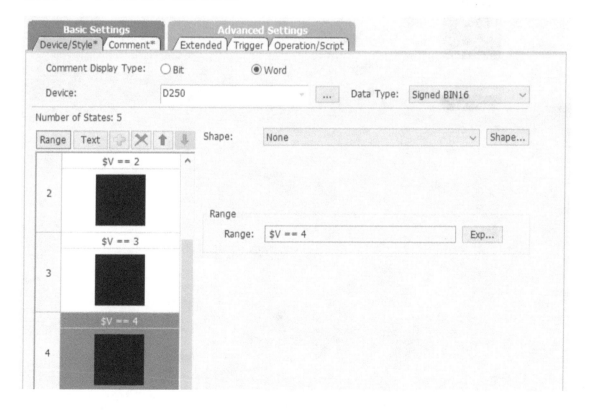

3. First, select condition 0. If D250=0, nothing is displayed, so select "Hold" for the comment display method. Also, set the Text Size to 26.

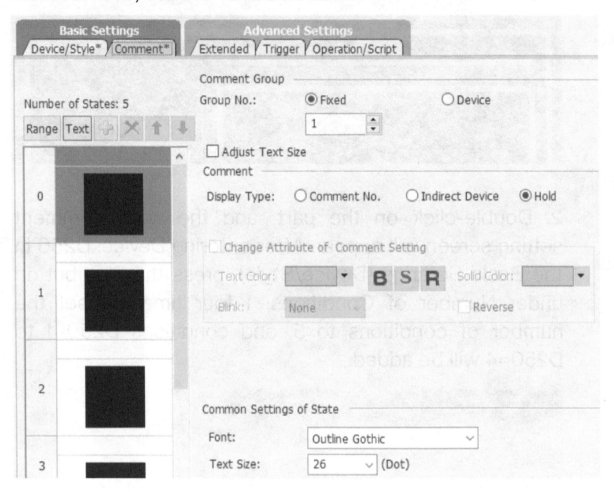

4. Next, select Condition 1, where D250=1, and since you want to display "Popcorn" leave the comment as "Comment No." and the comment No. as 1, press the "Edit..." button, enter "Popcorn" on the comment edit screen, and press "OK.

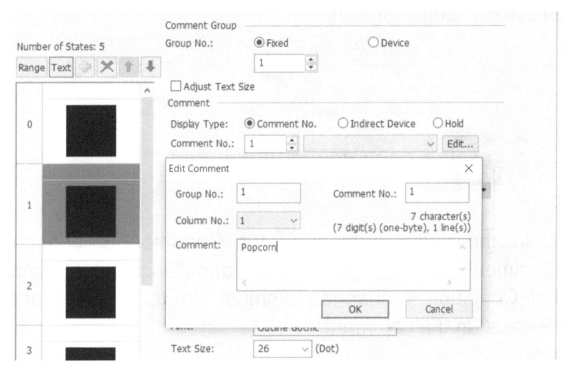

5. When the following message appears, press "Yes"(はい) to reflect the "Popcorn" in the comment list. This completes the setting of Condition 1.

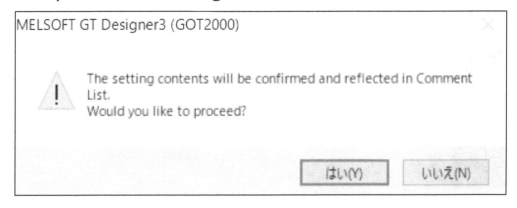

6. Next, select Condition 2, and in the case of D250=2, you want to display "Beverage" so leave the comment No. asit is, change the comment No. to 2, press the "Edit..." button, enter "Beverage" on the comment edit screen, and press "OK" or "Yes"(はい) to display the comment. Press the "Edit..." button.

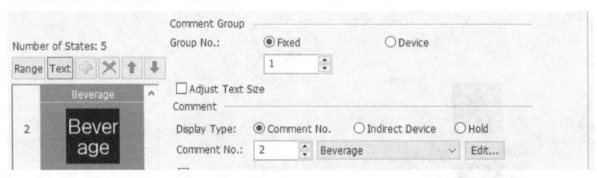

7. In the same procedure, for Condition 3, change Comment No. to 3, and reflect "Potato" in the comment. For Condition 4, change Comment No. to 4 and reflect "Reheat" in the comment.

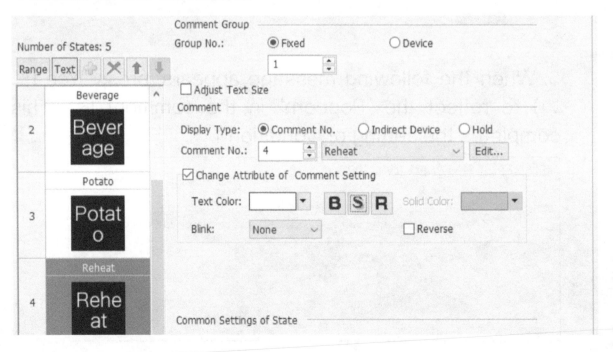

8. adjust the position and size of the part. (Displayed in the lower left corner after clicking on the part)
X:445 Y:70 W:150 H:40

9. create a button for each mode. First, create the "Range" button. After selecting "Object" tab ⇒Switch ⇒ Word Switch, click on the screen to place the parts.

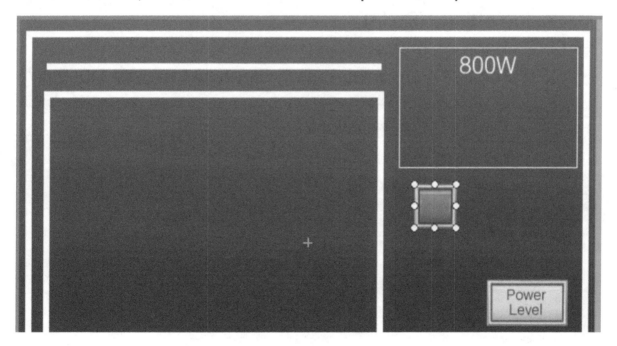

10. Double-click the component to display the word switch setting screen and make the following settings.
Basic Settings (Devices)
Device: D250   Constant:1

Basic Settings (Style)
OFF shape color: white
OFF shape: FA Control Switch Figure A1 57 SW_02_0_W

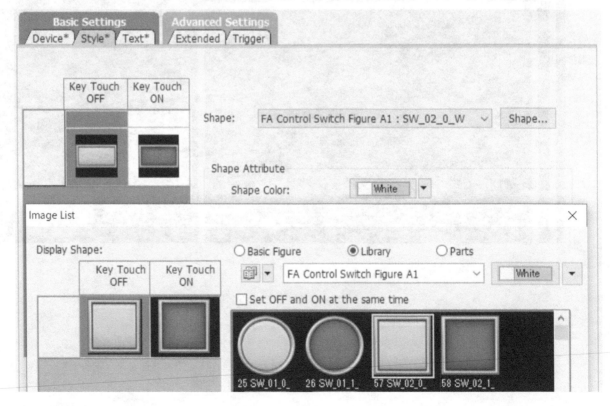

240

ON shape color: white
ON shape: FA Control Switch Figure A1 58 SW_02_1_W

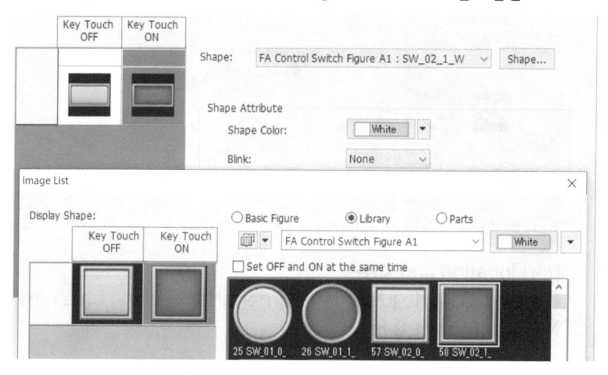

[Basic settings (text)]
Text type OFF=ON check: OFF
Text size: 16
Text OFF: Popcorn
Text color: black

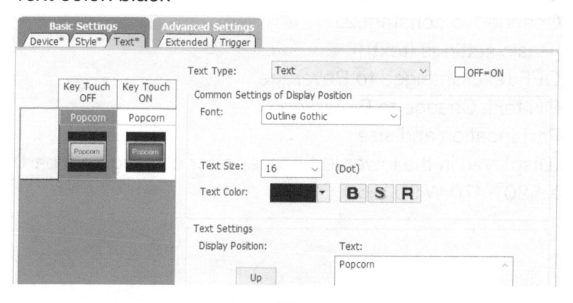

Text ON: Popcorn
Text color: white

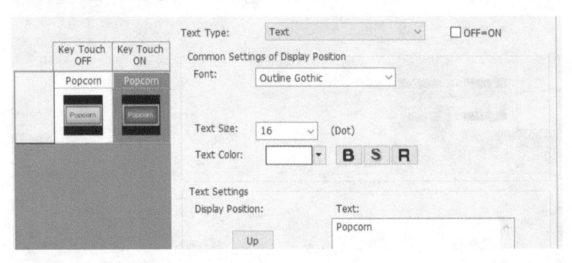

Part location and size
(Displayed in the lower left corner after clicking on a part)
X:420 Y:170 W:90 H:50

11. copy => paste the "Popcorn" button to create the "Beverage", "Potato", and "Reheat" buttons.

"Beverage" button.
Copy => Paste "Popcorn" button
Basic Settings (Devices)
Changed to constant:2.
[Basic settings (text)]
OFF text: Changed to Beverage
ON text: Change to Beverage
Part location and size
(Displayed in the lower left corner after clicking on a part)
X:520 Y:170 W:90 H:50

"Potato" button.
Copy => Paste "Popcorn" button
Basic Settings (Devices)
Changed to constant:3.
[Basic settings (text)]
OFF text: Change to Potato
ON text: Change to Potato
Part location and size
(Displayed in the lower left corner after clicking on a part)
X:420 Y:225 W:90 H:50

"Reheat" button.
Copy => Paste "Popcorn" button
Basic Settings (Devices)
Constant: Changed to 4.
[Basic settings (text)]
OFF text: changed to Reheat.
ON Text: changed to Reheat n.
Part location and size
(Displayed in the lower left corner after clicking on a part)
X:520 Y:225 W:90 H:50

12. After creating the screen, start the PLC program used in the past simulation, start the simulator for the screen, and confirm the following operations. (It is not necessary to create the PLC program.)

Pressing each mode button displays each mode on the monitor.

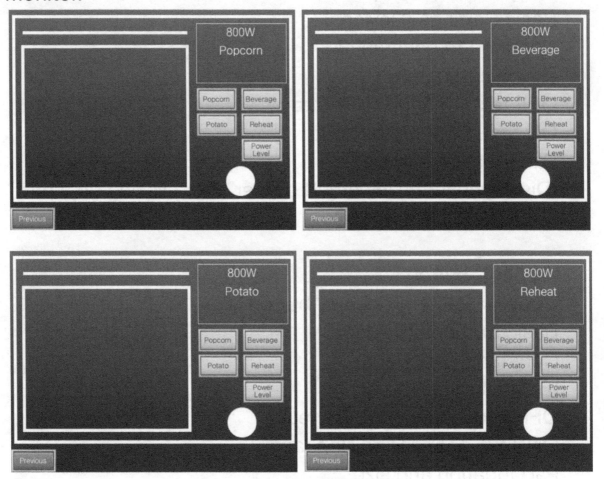

6-3 to 6-4 Summary

Bit comments are used to switch comments on and off, rather than switching the color of the lamp like bit lamps. Word comments can switch between multiple comments by the number in the specified device.

Exercise 9 (Watch out for frozen roads!)

Create a new screen number 33, question 9, and create a monitor screen that displays " Watch out for frozen roads! " When the current temperature (numerical input part: D260) drops below minus 3°C. (bit comment: Y0050) when the current temperature (numerical input part: D260) drops below minus 3°C. (Also add a button to move between screens 32 and 33)

PLC program for simulation (how to write is explained in the solution)

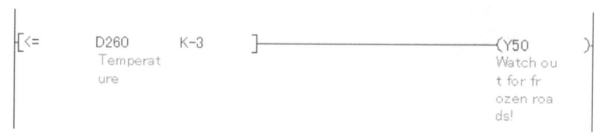

Exercise 9 Solution

Caution! Display

"Object" tab => Comment display ⇒Use bit comments

Basic Settings (Device/Style)

Device: Y0050

Comment Display Type:	◉ Bit	○ Word
Device:	Y0050	...

Basic Settings (Comment)

OFF=ON check: OFF   Display Type: Text

ON: Watch out for frozen roads!

Text color: Yellow Bold (Click B)

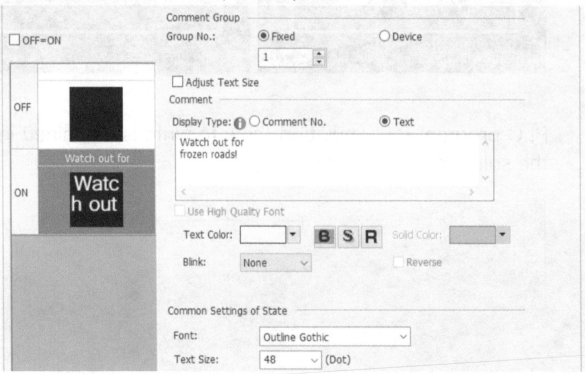

*After that, you are free to set up as you wish.

(Reference setting)
Basic Settings (Comment)
Text size: 48

Part location and size
(Displayed in the lower left corner after clicking on a part)
X:120 Y:100 Width:400 Height:100

Temperature display
"Object" tab => Numeric Display/Input
⇒Use numerical input

Basic Settings (Devices)
Device: D260

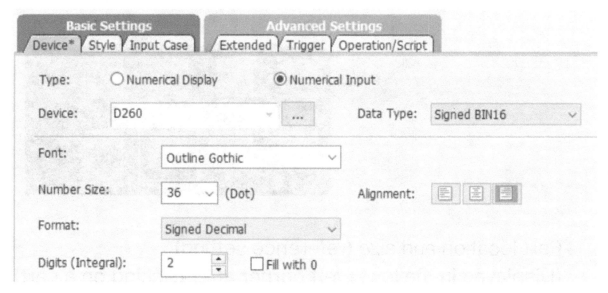

(Reference setting)
Basic Settings (Devices)
Numeric size: 36 Integer digits: 2
Show "+": Check
Preview value: 12

Basic Settings (Style)
Shape: Square_3D_Fixed_Width: Rect_7

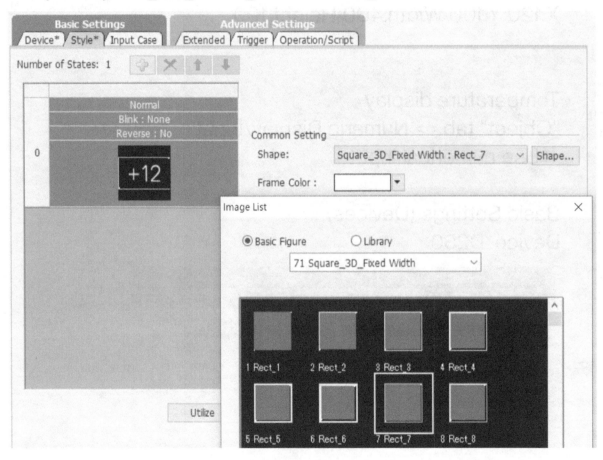

Part location and size (reference setting)
(Displayed in the lower left corner after clicking on a part)
X:330 Y:295 Width:85 Height:50

Monitor screen frame (reference setting)
Figure tab => Rectangle
Line width: 7Dot

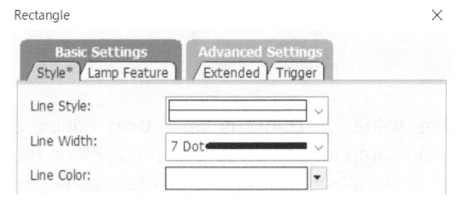

Figure position and size
(Displayed in the lower left corner after clicking on the shape)
X:65 Y:50 W:500 H:340

Text: temperature (reference setting)
Figure tab ⇒After selecting Texts, click on the screen
Text:temperature   Text size: 36
Figure position and size
(Displayed in the lower left corner after clicking on the shape)
X:120 Y:300 Width:180

Text: °C (reference setting)
Figure tab ⇒After selecting Texts, click on the screen
Text:°C   Text size:36
Figure position and size
(Displayed at the lower left after clicking on the shape)
X:440 Y:300 Width:36

PLC program for simulation

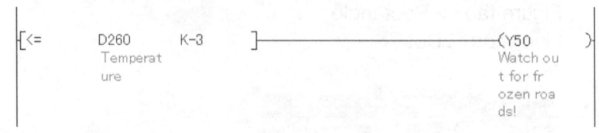

```
├─[<=     D260      K-3     ]─────────────────────────────(Y50        )┤
           Temperat                                        Watch ou
           ure                                             t for fr
                                                           ozen roa
                                                           ds!
```

ON when the value of D260 is less than minus 3 (numerical value comparison instruction): press F8 on the keyboard, entercircuit:<= d260 k-3, enter comment (D260): Temperature

Y50: press F7 on keyboard, enter circuit: enter Y50, enter comment: Watch out for frozen roads!

Finally, press F4 to convert. (If you cannot convert, see 4-10)

Now that the simulation preparation is finished, execute the simulation startup of GX Works2 and the simulator startup of GT Designer3 to confirm the following operations.

When you click on the current temperature display and enter minus 3 degrees Celsius or lower, "Watch out for frozen roads!" is displayed.

If you click on the current temperature display and enter a temperature greater than minus 3°C, nothing isdisplayed.

Exercise 10 (Washing machine status display monitor)

Create a monitor (word comment component: D270) that displays "washing" at 26 to 45 minutes remaining time (numeric input component: D270), "rinsing" at 11 to 25 minutes remaining time, "drying" at 01 to 10 minutes remaining time, and "completion" at 00 minutes remaining time on screen number 30, question 7. (PLC program for simulation is not required.)

[Reference] Setting the range of "washing".

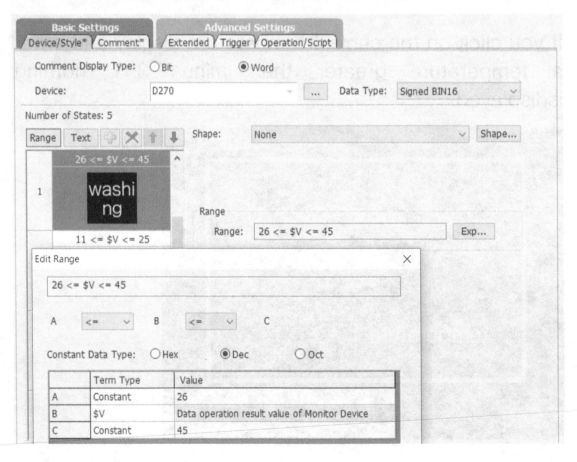

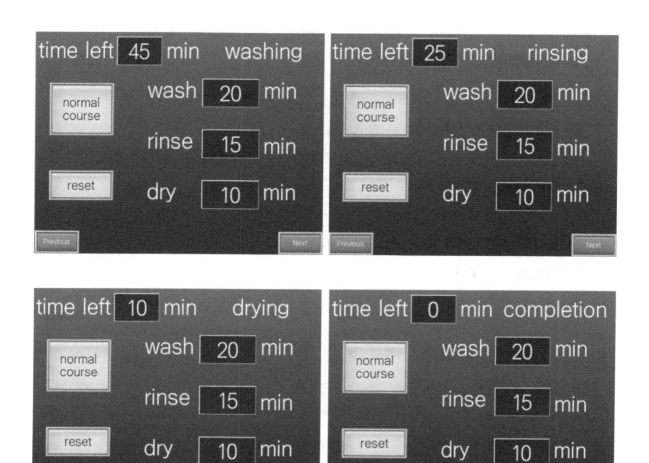

Exercise 10 Solution

Washing Condition Indication
"Object" tab => Comment display ⇒Use word comment

<Condition 0>.
Basic Settings (Device/Style)
Device: D270

Word Comment Display

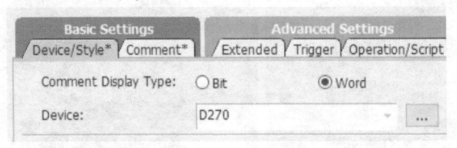

Basic Settings (Display Comments)
Comments Display method: No treatment

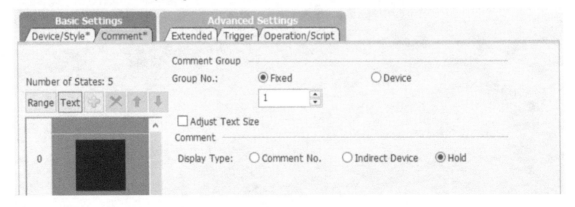

*After that, you are free to set up as you wish.
(Reference setting)
Basic Settings (Display Comments)
Text size: 48

Condition 1.

Basic Settings (Device/Style)

Plus, mark below the number of conditions.

After pressing and adding a condition,

Click on "Exp..." for the range.   26 ≤D270($V) ≤45

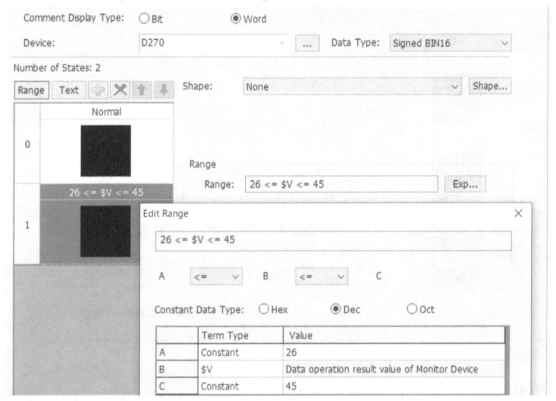

Basic Settings (Display Comments)

Set comment No. to 10 and set "washing".

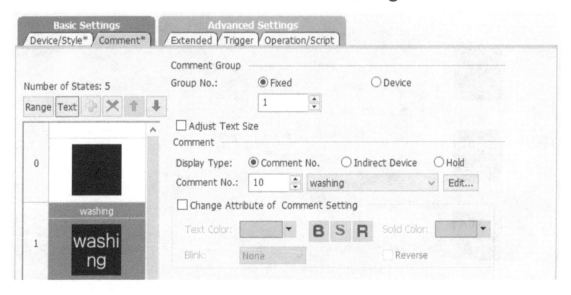

Condition 2.
Basic Settings (Device/Style)
Plus, mark below the number of conditions.
After pressing and adding a condition,
Click on "Exp..." for the range.   11≤D270($V) ≤25

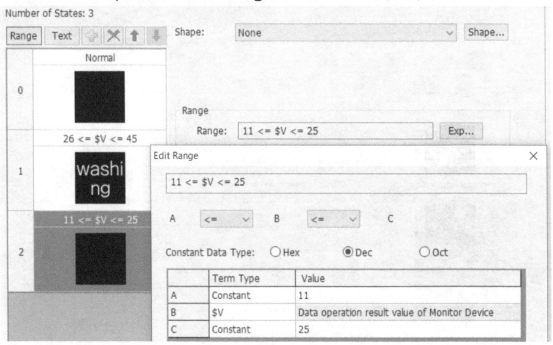

Basic Settings (Display Comments)
Set "rinsing" with comment no. 11.

Condition 3.

Basic Settings (Device/Style)

Plus, mark below the number of conditions.

After pressing to add a condition,

Click on "Exp..." for the range. $1 \leq D270(\$V) \leq 10$

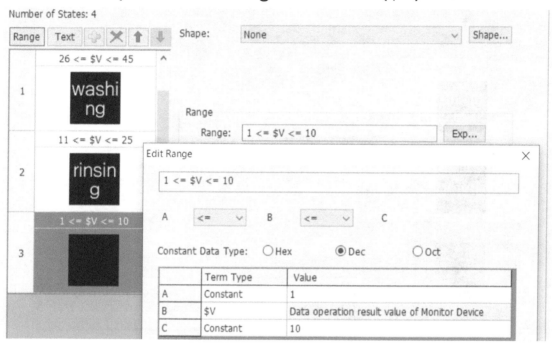

Basic Settings (Display Comments)

Set "drying" with comment no. 12.

Condition 4.

Basic Settings (Device/Style)

Plus, mark below the number of conditions.

After pressing to add a condition,

Click on "Range..." for the range.   D270($V) ==0

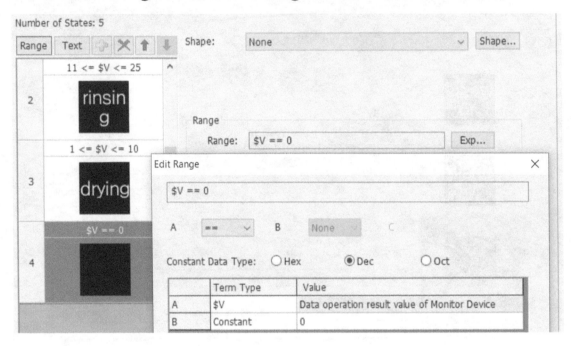

Basic Settings (Display Comments)

Set "completion" with comment no. 13.

Part location and size (reference setting)
(Displayed in the lower left corner after clicking on a part)
X:380 Y:15 W:250 H:50

Text: time left (reference setting)
Figure tab ⇒After selecting Texts, click on the screen
Text: time left   Text size:48
Figure position and size
(Displayed at the lower left after clicking on the shape)
X:5 Y:15 Width:168

Text: min (reference setting)
Figure tab ⇒After selecting Texts, click on the screen
Text: min   Text size: 48
Figure position and size
(Displayed in the lower left after clicking on the shape)
X:290 Y:15 Width:74

Time left input field
Wash time indicator component.
Copy ⇒Paste

Basic Settings (Device/Style)
Type: Changed to Numerical Input.
Device: Changed to D270

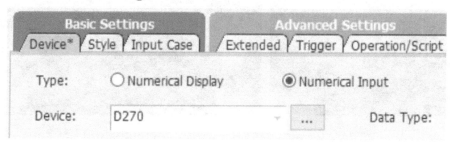

When you have finished creating the screen, start the PLC program used in the past simulation, start the simulator for the screen, and confirm the following operations. (It is not necessary to create the PLC program.)

After pressing the standard course button, click the remaining machine time and enter 26 to 45 minutes to display "washing", 11 to 25 minutes to display "rinsing", 01 to 10 minutes to display "drying", and 00 minutes to display "completion".

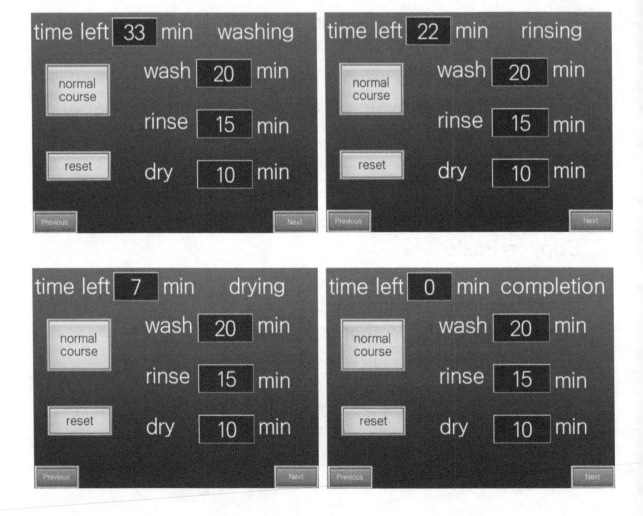

6-5 Date/Time Display

The date and time can be displayed on the touch panel by using thedate/time display component in the object.
This time, we will create thedate and time in the upper right corner of the Screen 10: Switch Lamp screen.

1. First, create a date. Open Screen 10, select the "Object" tab ⇒Date/Time Display ⇒Date Display, then click on the screen.

2. Double-click the component to display the date setting screen and make the following settings. There are various date display formats, but we will leave the default settings as they are.

Basic Settings (Text/Style)
Text size: 24

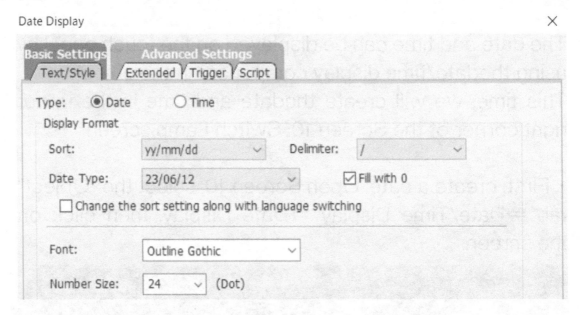

Part location and size
(Displayed in the lower left corner after clicking on a part)
X:440 Y:5 Width:96

3. Next, create the time. Select "Object" tab ⇒Date/Time
Display ⇒Time Display and click on the screen.

4. Double-click the component to display the time setting screen and make the following settings.
Basic Settings (Text/Style)
Time format: hours: minutes: seconds
Text size: 24

Part location and size
(Displayed inthe lower left corner after clicking on a part)
X:540 Y:5 Width:96

After creating the screen, start the PLC program used in the past simulations, start the simulator for the screen, and confirm the following operations. (It is not necessary to create the PLC program.)

The correct date, time, and date/time are displayed in the upper right corner of the screen and the time is updated on the screen.

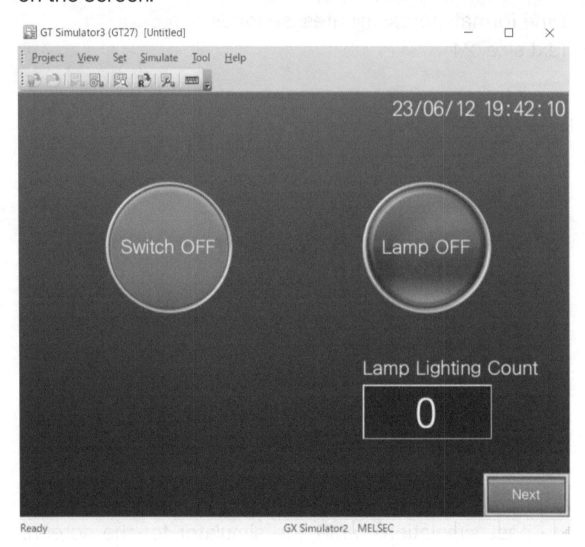

6-5 Summary

The date and time can be displayed on the touch panel by using the date/time display component.

Exercise 11 (date and time display)

In the upper left corner of Screen 11 (Question 1), put the date and time in the display format of the following screen.

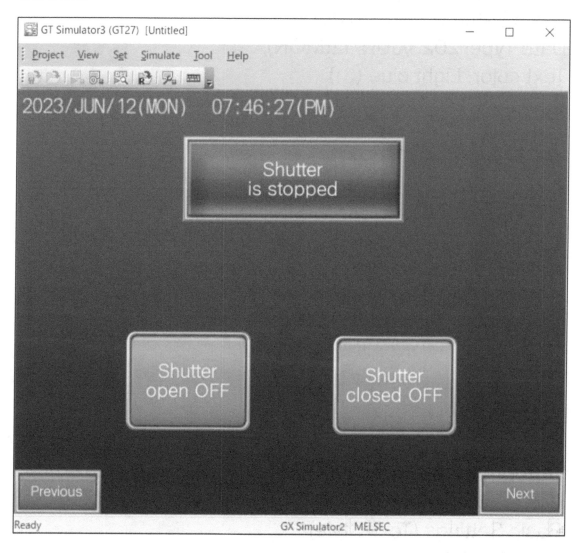

Exercise 11 Solution

Date display
"Object" tab => Date/Time Display
⇒Use date display
Basic Settings (Text/Style)
Date Type: 2023/JUN/12(MON)
Text color: Light blue (31)

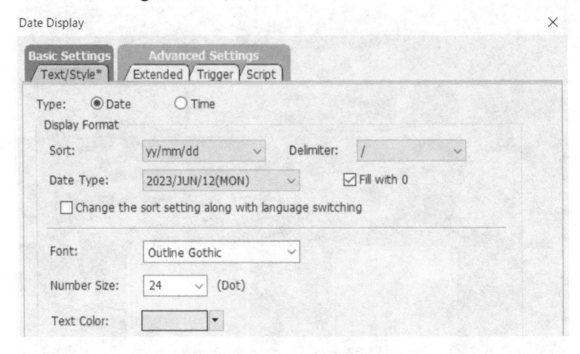

*After that, you are free to set up as you wish.
(Reference setting)
Basic Settings (Text/Style)
Text size: 24
Part location and size (reference setting)
(Displayed in the lower left corner after clicking on a part)
X:5 Y:5 Width:192

Time display
"Object" tab => Date/Time Display
⇒Use time display
Basic Settings (Text/Style)
Time Type: 07:52:40(PM)
Text color: Light blue (31)

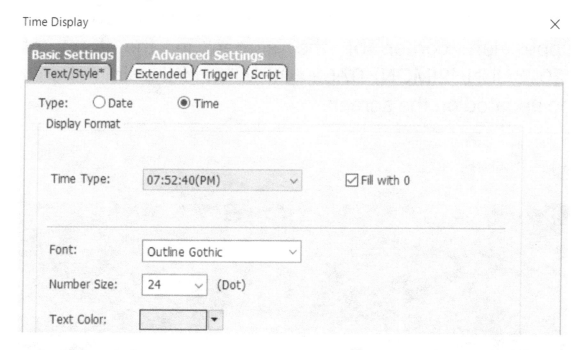

(Reference setting)
Basic Settings (Text/Style)
Text size: 24
Part location and size (reference setting)
(Displayed in the lower left corner after clicking on a part)
X:225 Y:5 Width:144

When you have finished creating the screen, start the PLC program used in the past simulations, start the simulator for the screen, and confirm the following operations. (It is not necessary to create the PLC program.)

The correct date and time should be displayed in the upper left corner of the screen in the form of "2023/JUN/12(MON) 07:52:40(PM)" and the time should be updated on the screen.

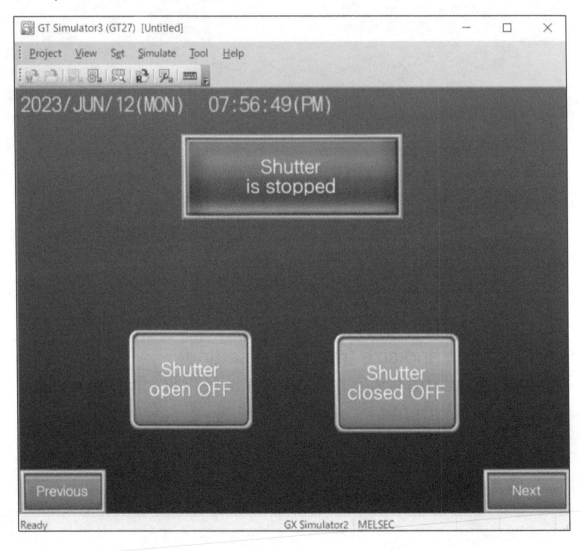

Chapter7 Other Objects and Functions

7-1 Screen Call

Screen recall is the ability to recall a created screen from another screen.

It is OK to copy and paste a group of parts or figures to multiple screens without using screen calls, but if revisions are necessary, copying and pasting will require revisions on each screen. In contrast, with screen calls, only the screen being called needs to be modified, which saves time.

This time, we will use screen calls to create a move button on each screen to go to the respective screen. We will also make sure that the move button for the screen being displayed is yellow.

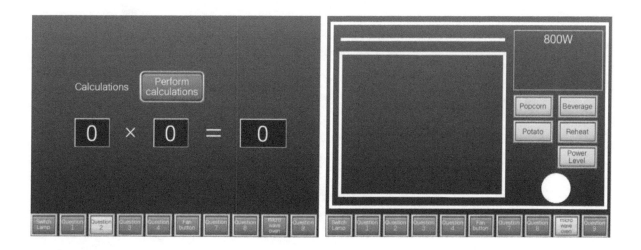

1. first, create the original screen from which the screen will be called. Create a new screen with screen number: 50 and title: Moving Between Screens.

Screen Property

Basic	Key Window Basic Setting	Key Window Advanced Setting

Screen No.: 50

Screen Name: Moving Between Screens

2. select "Object" tab ⇒switch ⇒Go To Screen Switch, then click on the screen.

3. Double-click the part to display the screen switching switch setting screen and make the following settings.

Basic Settings (Next Screen)
Screen number:10   Lamp function: Word Range
Device: D1000   ON range:10(screen number) ==D1000

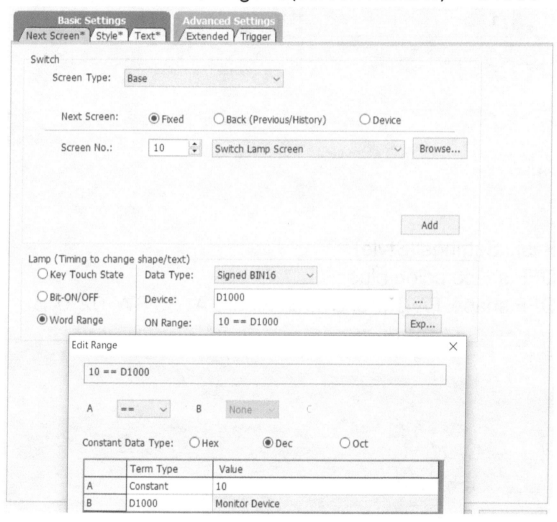

The D1000 being set here is the device that was set when the project was newly created. This D1000 will contain the number of the currently displayed screen.

This setting can be found in the "Common " tab ⇒GOT Environmental Setting ⇒Screen Switching/Windows.

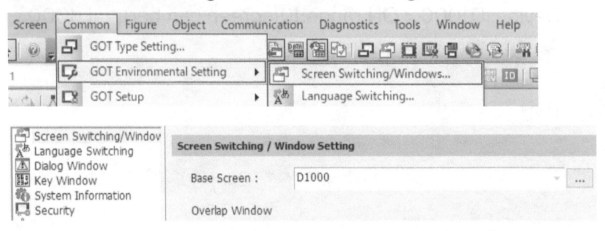

Basic Settings (Style)
OFF shape color: blue
OFF shape: FA Cntrol Switch Figure A1 33 SW_02_0_B

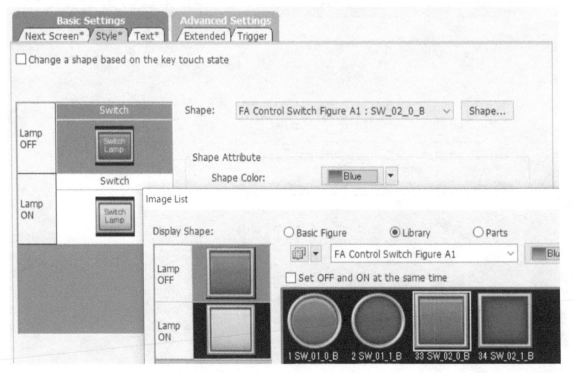

ON shape color: yellowish
ON shape: FA Control Switch Figure A1 37 SW_02_0_Y

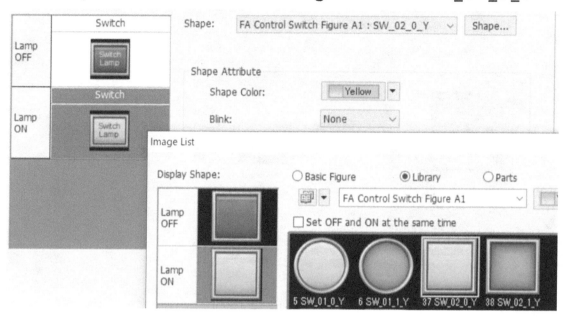

[Basic settings (text)]
Text type OFF=ON check: OFF
Text size:12
Text OFF: Switch Lamp
Text color: white

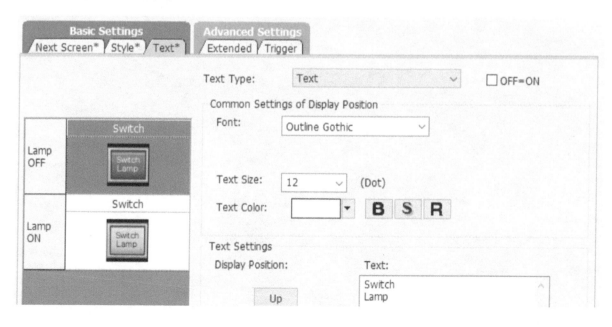

Text ON: Switch Lamp
Text color: black

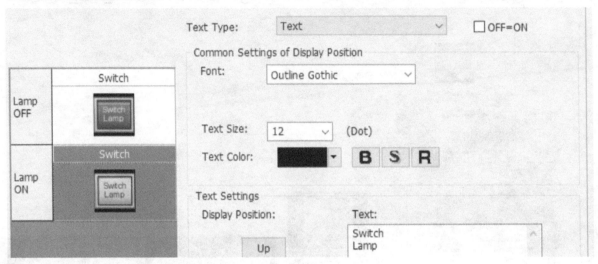

Part location and size
(Displayed in the lower left corner after clicking on a part)
X:0 Y:430 Width:60 Height:50

4. Right-click on the created part ⇒Select "Consecutive Copy" and set the number of the copied parts as follows: X direction: 10, Y direction: 1, Spacing X direction: 5, Increment target: All check OFF, and press "OK".

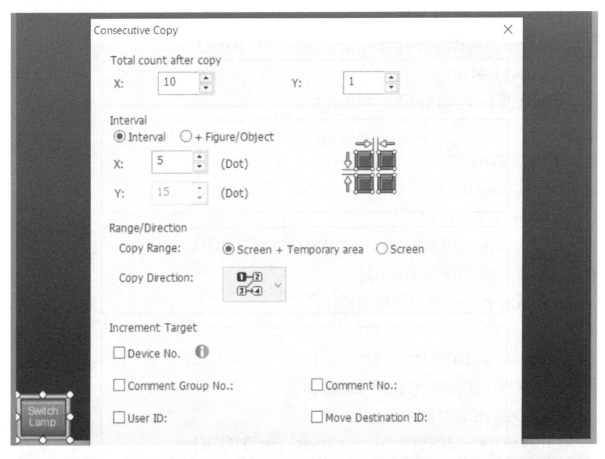

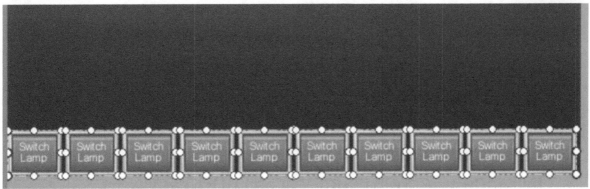

5. set each screen switching switch.

Second switch from the left
Basic Settings (Next Screen)
Screen number:11
ON range:11(screen number) ==D1000
[Basic settings (text)]
ON/OFF Texts: Question 1

Third switch from the left
Basic Settings (Next Screen)
Screen number:12
ON range:12(screen number) ==D1000
[Basic settings (text)]
ON/OFF Texts: Question 2

Fourth switch from the left
Basic Settings (Next Screen)
Screen number:13
ON range:13(screen number) ==D1000
[Basic settings (text)]
ON/OFF Texts: Question 3

Fifth switch from the left
Basic Settings (Next Screen)
Screen number:14
ON range:14(screen number) ==D1000
[Basic settings (text)]
ON/OFF Texts: Question 4

Sixth switch from the left
Basic Settings (Next Screen)
Screen number:22
ON range:22(screen number) ==D1000
[Basic settings (text)]
ON/OFF Text: Fan button

Seventh switch from the left
Basic Settings (Next Screen)
Screen number:30
ON range:30(screen number) ==D1000
[Basic settings (text)]
ON/OFF Text: Question 7

Eighth switch from the left
Basic Settings (Next Screen)
Screen number:31
ON range:31(screen number) ==D1000
[Basic settings (text)]
ON/OFF Text: Question 8

9th switch from the left
Basic Settings (Next Screen)
Screen number:32
ON range:32(screen number) ==D1000
[Basic settings (text)]
ON/OFF text: Microwave oven

Tenth switch from the left
Basic Settings (Next Screen)
Screen number:33
ON range:33(screen number) ==D1000
[Basic settings (text)]
ON/OFF Text: Question 9

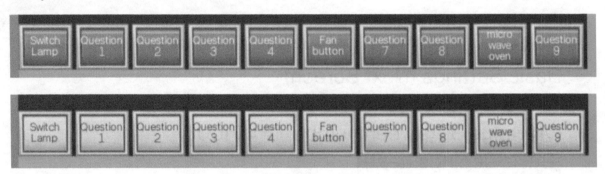

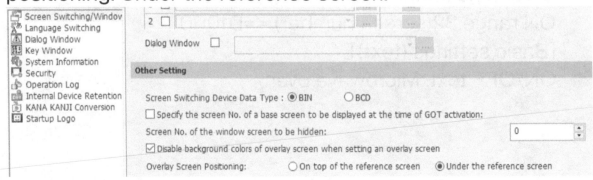

6. Now that the screen is completed, the necessary settings must be made before the screen calls can bemade on each screen. Open "Common" tab ⇒GOT Environmental Setting ⇒ Screen Switching/Window, scroll down the screen, make the following settings and press "Apply" and "OK".

Other setting
Disable background colors of overlay screen when setting an overlay screen: check, Overlay screen positioning: Under the reference screen.

7. open screen 10 and delete the "Next" button. Then select "Object" tab ⇒ Set Overlay Screen, select the screen 50 you wish to recall and press "OK".

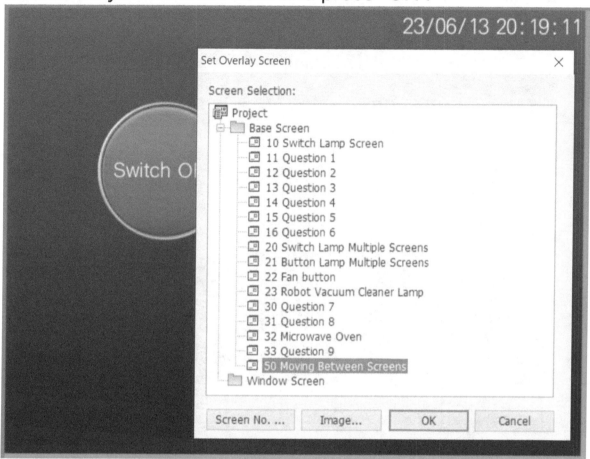

8. Click on the screen to call up Screen 50, and while left clicking, bring it to the upper left or set XY in the lower left to X:0 Y:0.

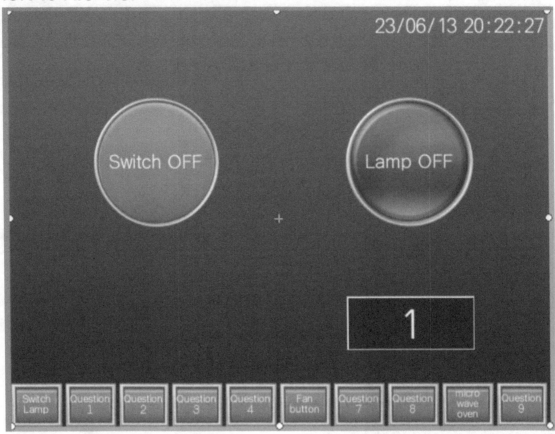

9. call up the remaining 9 screens using the same procedure as Screen 10. You can also copy and paste the screen call from Screen 10.

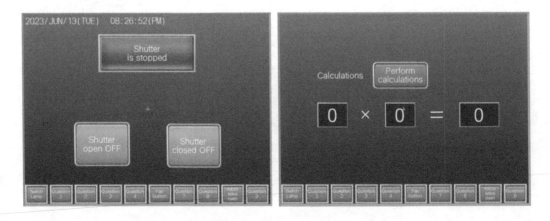

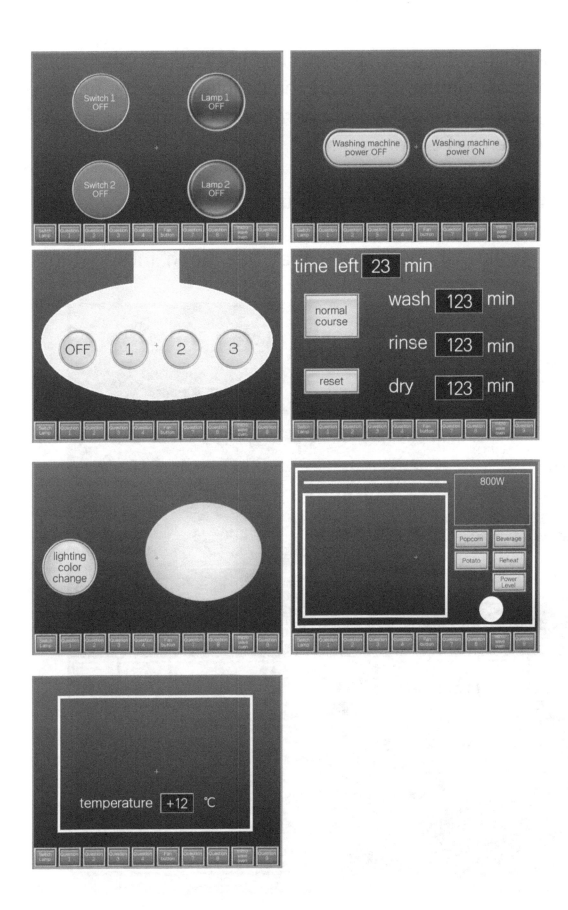

10. After creating the screen, start the PLC program used in the past simulations, start the simulator for the screen, and confirm the following operations. (It is not necessary to create the PLC program.)

Can navigate to each screen using the toggle switch for screen recall.
The switch of the screen that has been moved turns yellow.

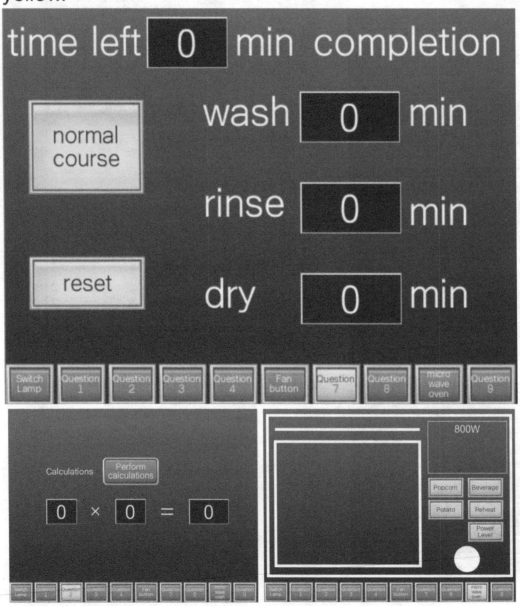

7-2 Window Screen

A window screen is ascreen that can be displayed as a pop-up screen on the base screen. The "Operation Explanation" function displays a window screen to explain the operation whenthe HELP button is pressed in an area where the operation is difficult to understand, and the "Operation Confirmation" function displays a window screen when an operation button is pressed to confirm whether to execute the operation. Pressing "No" will cancel the operation.

This time, I will go ahead and add the Show Description button to Screen 31 (Question 8) and create a function that, when pressed, will display a window screen, and show the description.

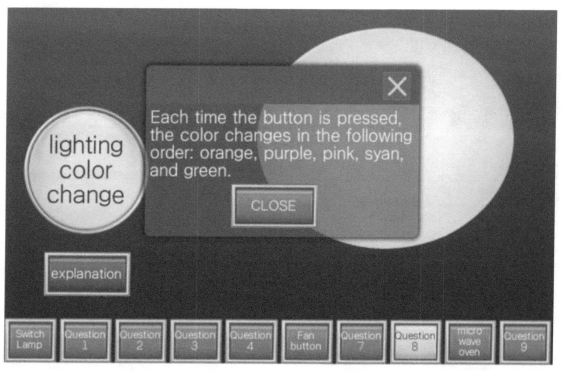

1. first create a window screen. Double-click New Window Screen, enter Screen Number: 1, Title: Button Explanation, and press "OK.

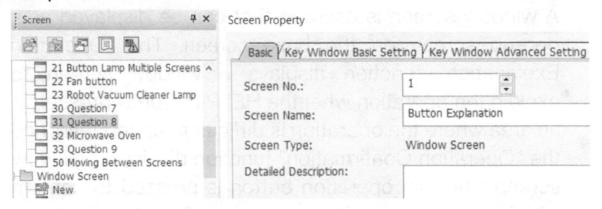

2. create each component in window screen 1.
Comment
"Figure" tab => Use text
Text: Each time the button is pressed, the color changes in the following order: orange, purple, pink, syan, and green.
Text size: 20

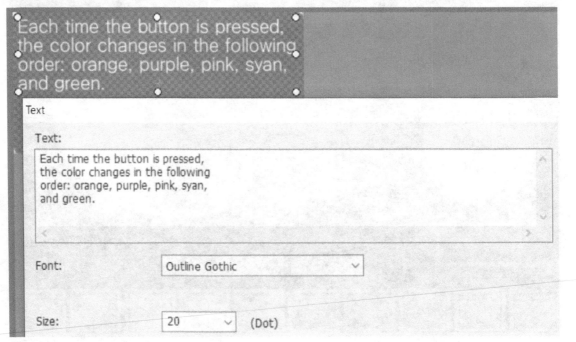

Figure position and size
(Displayed in the lower left corner after clicking on a part)
X:15 Y:20 Width:280

"CLOSE" button
Object" tab => Switch ⇒Use switch parts

Basic Settings (Action)
After selecting the screen switching for additional operation, set the Screen type: Overlap Window1, screen number: 0 (close the window).

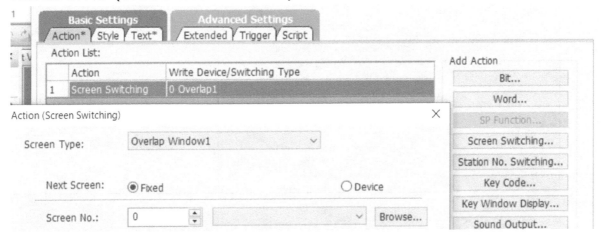

[Basic settings (text)]
ON/OFF text: CLOSE

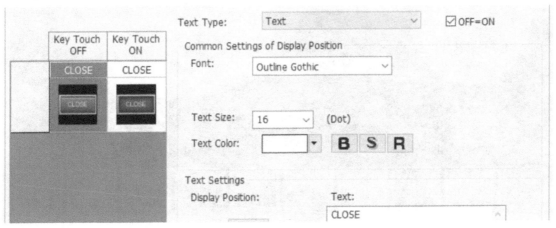

Part location and size
(Displayed in the lower left corner after clicking on a part)
X:100 Y:90 W:100 H:50

3. Right-click on the screen ⇒Select Screen Properties, click on "Screen Size..." And change the window screen size Y to 150. The window screen is now complete.

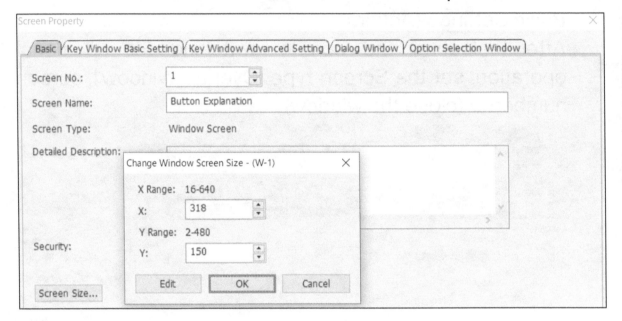

4. add a button to screen 31 (problem 8) to display the window screen. After selecting the "Object" tab ⇒Switch ⇒Go to Screen switch, click on the screen.

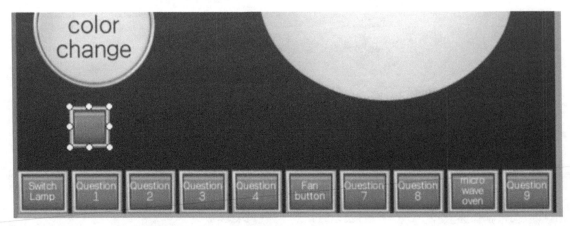

5. Double-click the part to display the screen switching switch setting screen and make the following settings.
Basic Settings (Next Screen)
Switching screen type: Overlap window1.
Screen number:1

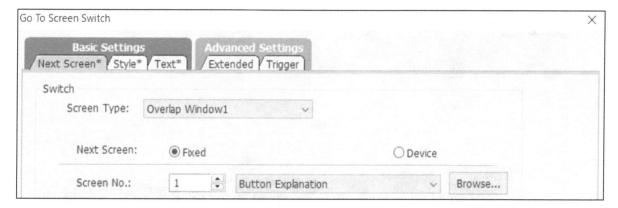

Basic Settings (Style)
OFF shape color: black
OFF shape: FA Control Switch Figure A1 55 SW_02_0_B

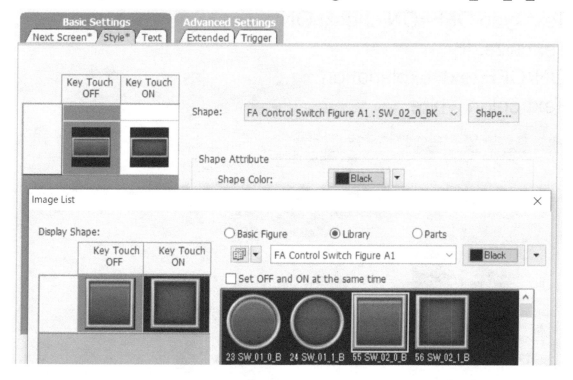

ON shape color: black
ON shape: FA Control Switch Figure A1 56 SW_02_1_B

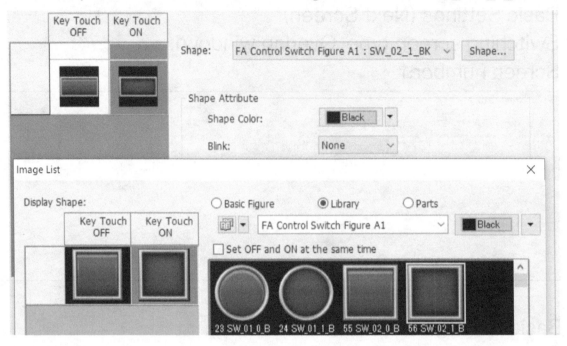

[Basic settings (text)]
Text type OFF=ON check: ON
Text size: 16
ON/OFF text: explanation
Text color: white

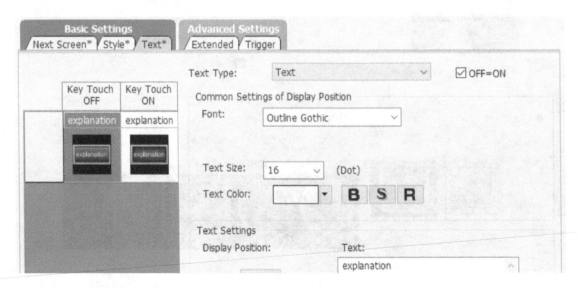

Part location and size
(Displayed in the lower left corner after clicking on a part)
X:45 Y:355 Width:100 Height:50

6. After creating the screen, start the PLC program used in the past simulation, start the simulator for the screen, and confirm the following operations. (It is not necessary to create the PLC program.)

Pressing the "explanation" button displays an explanation of the lighting color change.
Pressing the "Close" button or the"X" button in the upper right corner closes the window screen.

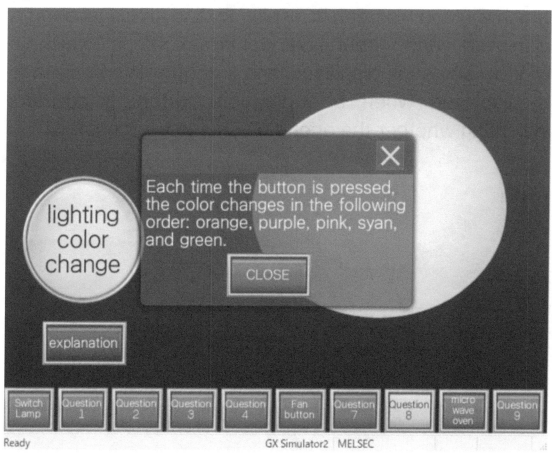

7-1 to 7-2 Summary

Screen recall is the ability to recall a created screen from another screen.

With screen recall, you only need to modify the screen you are recalling, which saves time.

A window screen is a screen that can be displayed as a pop-up screen on the base screen.

Window screen applications include "explanation" in which a window screen is displayed, and an explanation is displayed, and "operation confirmation" in which a window screen is displayed, and a confirmation message is displayed when an operation button is pressed, indicating whether the operation should be executed.

Exercise 12 (date/time display call)

Display the date and timecreated in Screen 11 (Question 1) on screens 1, 2, 3 and 4 using a screen callout. The screen to be called up should be the newly created screen number: 51 with the title: Date/Time Display.

Also, after calling to the screen for questions 1, 2, 3, and 4, change the time format on Screen 51 from 00:00:00 to 00:00:00.

Exercise 12 Solution

1. create a new screen number: 51, title: Date/Time Display, and copy/paste the date/timein the upper left corner of screen 11.

2. In screens 11, 12, 13, and 14, select Screen 51 in the "Object" tab ⇒ Set Overlay Screen and click on the screen to call up Screen 51.

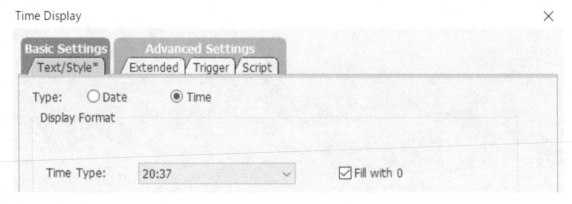

3. Double-click the time on Screen 51 and change the time format from 00:00:00(PM) to 00:00

Time Display   ✕

| Basic Settings | Advanced Settings |
| Text/Style* | Extended / Trigger / Script |

Type: ○ Date   ● Time
Display Format

Time Type: [20:37   ⌄]   ☑ Fill with 0

After creating the screen, start the simulation with the PLC program used in the past simulations, start the simulator for the screen, and confirm the following operations. (It is not necessary to create the PLC program.)

The current date and time shall be displayed in the upper left corner of screens 11, 12, 13, and 14, and shall be displayed in the format of 00:00:00(PM) to 00:00.

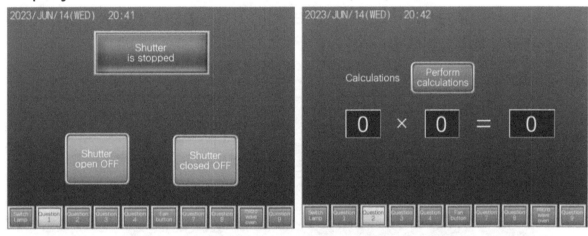

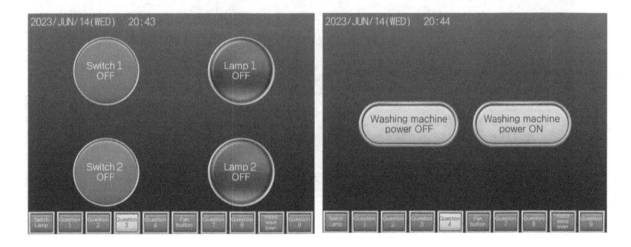

Exercise 13 (display of calculation execution confirmation screen)

When the Perform Calculation button on screen number 12 (Question 2) is pressed, window screen 2 (Confirm Perform Calculation) is displayed with the question "Do you want to run the calculation?" under "YES" to run the calculation and "NO" or the X in the upper right corner to cancel the calculation.

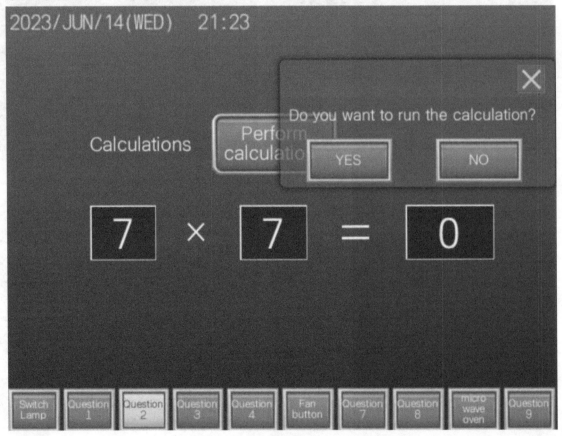

If "Yes" is pressed

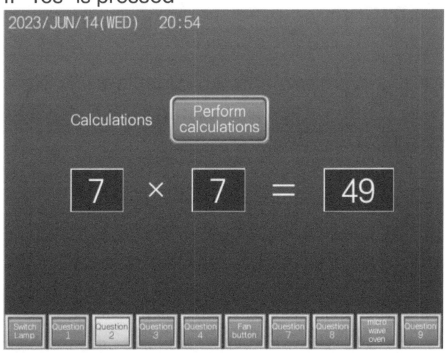

No" or press the "X" in the upper right corner.

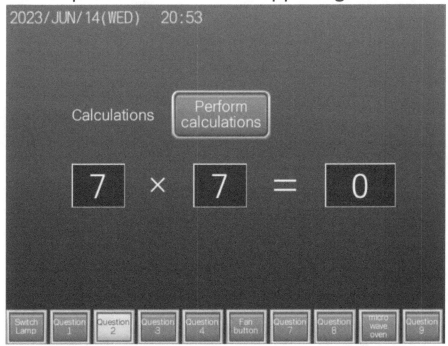

Exercise 13 Solution

1. First, create a window screen. Double-click New Window Screen, enter screen number: 2, title: Execution Confirmation, and press "OK.

Screen Property

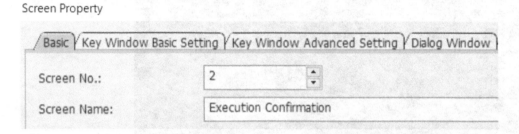

2. create each component on window screen 2.

Comments
"Figure" tab => Use text
Text: Do you want to run the calculation?
Text size: 18

Text

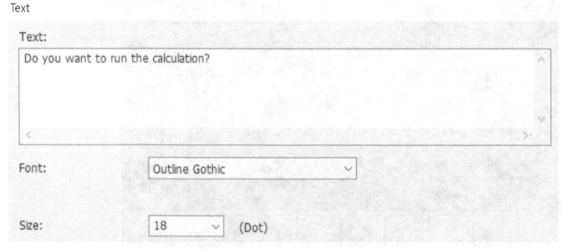

Figure position and size (reference setting)
(Displayed in the lower left corner after clicking on a part)
X:20 Y:5 Width:280

"Yes" button.

Object" tab => Switch ⇒ Switch

Basic Settings (Action)

After selecting additional bits of operation,

Device: X0003

Action: Bit Momentary

After selecting the screen switching for additional operation,

Switching screentype: Overlap window 1.

Screen number: 0 (close window)

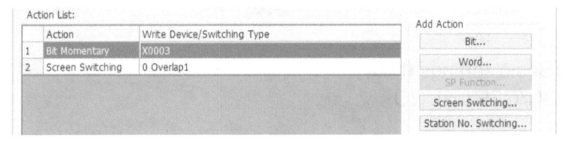

[Basic settings (text)]
ON/OFF text: YES

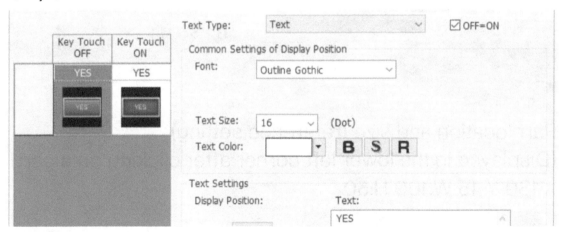

Part location and size (reference setting)
(Displayed in the lower left corner after clicking on a part)
X:30 Y:45 W:100 H:50

"No" button.
Copy => Paste "YES" button
Basic Settings (Action)
Device: X0003
Action: Momentary under Add Behavior after selecting press "Delete" to delete.

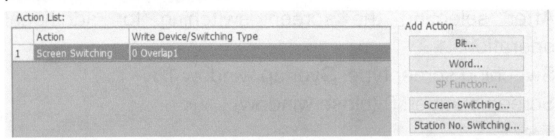

[Basic settings (text)]
ON/OFF text: NO

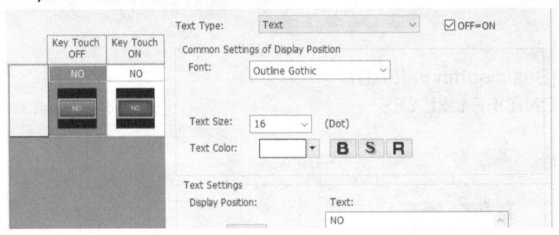

Part location and size (reference setting)
(Displayed in the lower left corner after clicking on a part)
X:180 Y:45 W:100 H:50

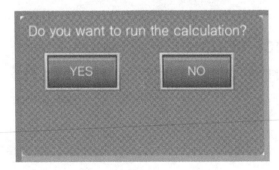

3. Right-click on the screen ⇒Select Screen Properties, click on "Screen Size..." And change the window screen size Y to 100. Window Screen 2 is now complete.

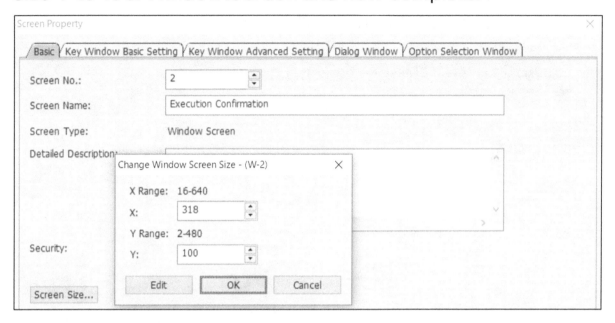

4. double-click on the Calculation Implementation button on Screen 12 to change the settings. First, after pressing the "Add Movement" button in the Movement Settings, select "Yes"(はい) to convert to a switch.

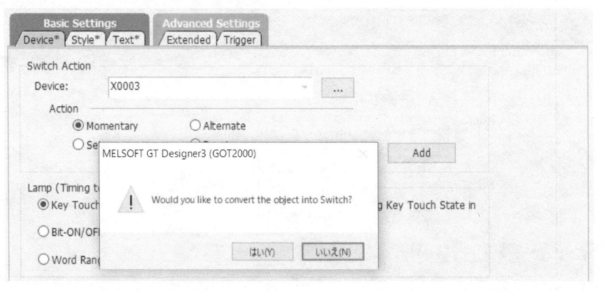

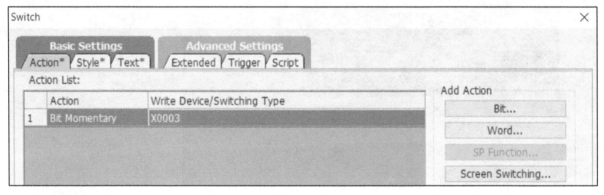

5. set the switch.

Basic Settings (Action)

Device: X0003

Action: After selecting bit momentary,

Press "Delete" under Add Behavior to remove it.

After selecting the screen switching for additional operation,

Switching screen type: Overlap window 1.

Set to screen number:2.

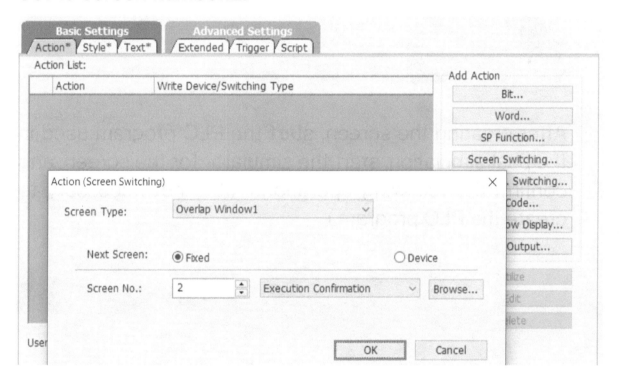

[Basic settings (text)]
Text type OFF=ON check: ON

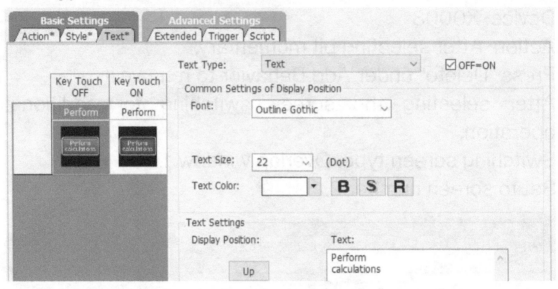

After creating the screen, start the PLC program used in the past simulation, start the simulator for the screen, and confirm the following operations. (It is not necessary to create the PLC program.)

After clicking the "Perform Calculation" button, the window "Do you want to run the calculation?" window will appear.

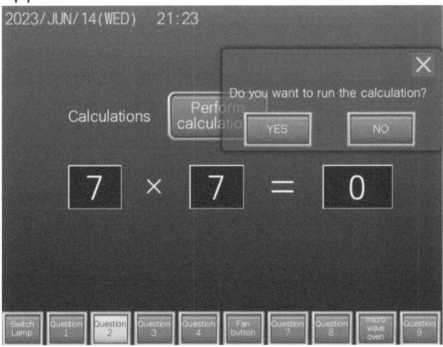

Pressing "YES" on the window screen closes the window screen and executes the calculation. If "NO" or the X in the upper right corner is pressed in the window screen, the window screen is closed, and the calculation is not executed.

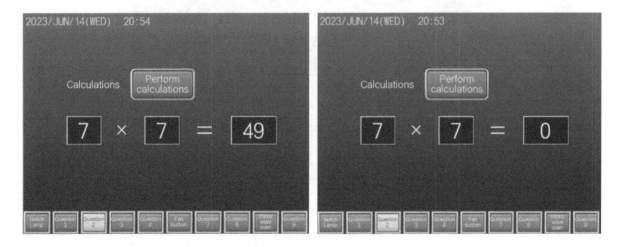

7-3 Display of Alarms Occurring

The Alarms in progress display allows you to check the alarms that are currently occurring when there is an abnormality in the facility, for example. The use is not for alarm history, but to quickly check only the alarms that are currently occurring.

This time, we will create a switch/lamp normal/failure switch button and an alarm display on the switch lamp screen of Screen 10 to check the display when an alarm occurs. (Originally, a failure signal is sent from the equipment, but this time it will be done with a button.) Switch/lamp both normal

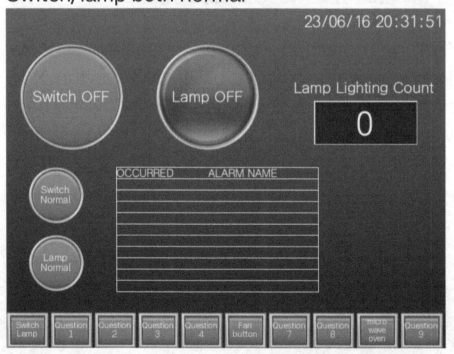

Failure of both switches/lamps

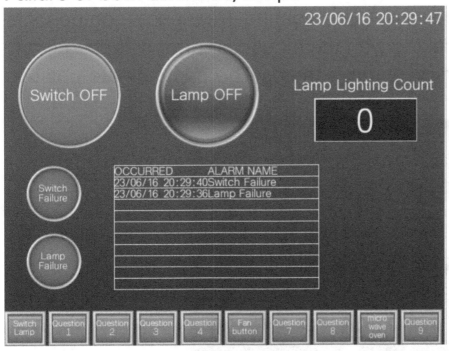

Failure of switch only

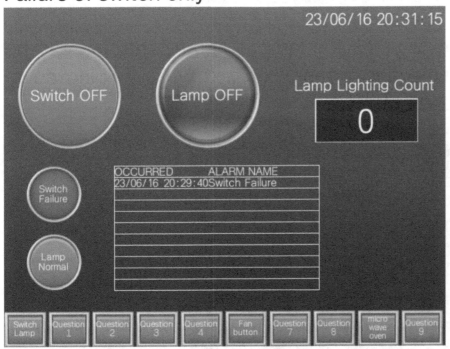

1. First, click "View" tab => Grid => Back, then move the switch, lamp, and lamp lighting frequency components while left clicking with the mouse, or after selecting them, use the cursor to move them. (The approximate position is OK.)

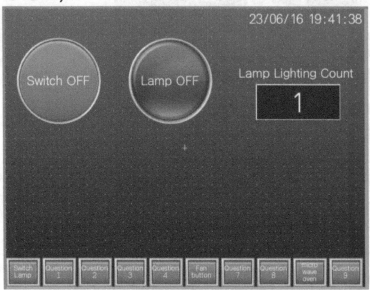

The number of times the lamp is lit, and the text can be moved by selecting a range of text while holding down the left mouse button.

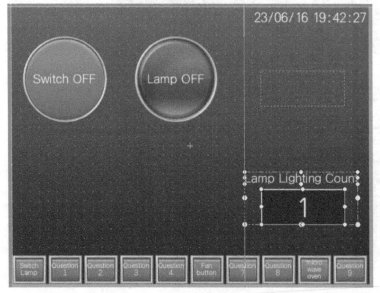

(Reference setting)
Switch: X:20 Y:50 Width: 150 Height: 150
Lamp: X:220 Y:50 Width:150 Height:150
Number of lights: X:450Y:130 Width:150 Height:62
String:X:420 Y:100 Width:203
*Strings can be selected by holding down Ctrl and clicking twice.

2. Next, create a switch/lamp normal/failure toggle button.
Button for switching between normal and faulty switches.
Object" tab => Switch ⇒ Bit switch.

Basic Settings (Action)
Device: Y0100
Action: Altemate
Lamp function: Bit ON/OFF
Lamp function device: Y0100

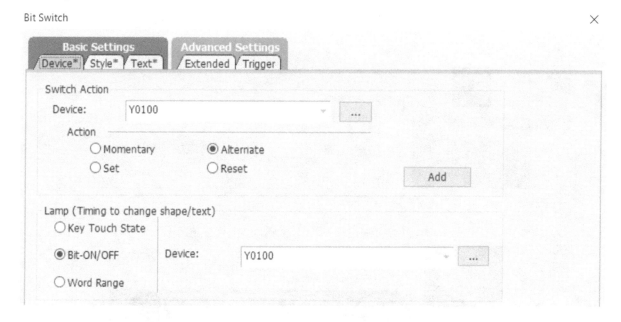

Basic Settings (Style)
OFF shape color: Blue
OFF shape: FA Control Switch Figure A1 1 SW_01_0_B

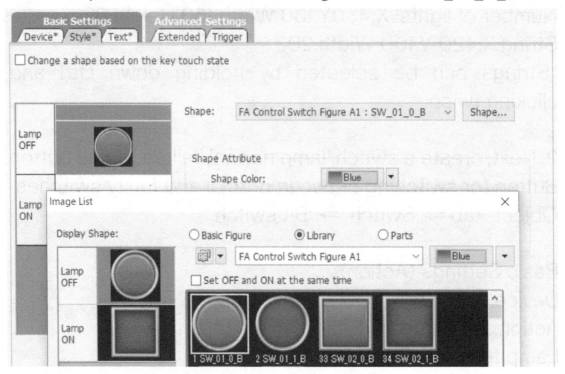

ON shape color: Red
ON shape: FA Control Switch Figure A1 4 SW_01_1_R

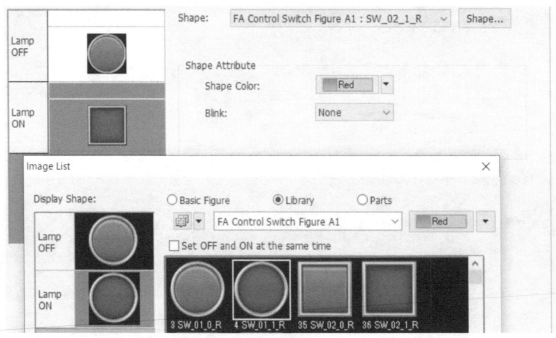

[Basic settings (text)]
Text type OFF=ON check: OFF
Text OFF: Switch Normal.
Text size: 14

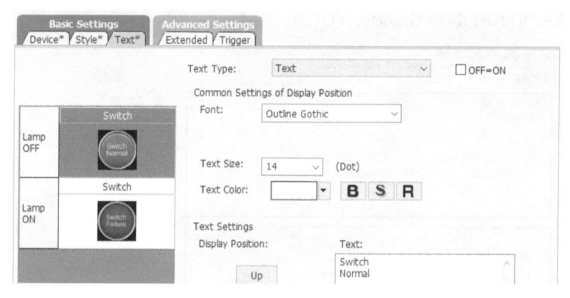

Text ON: Switch Failure
Text size: 14

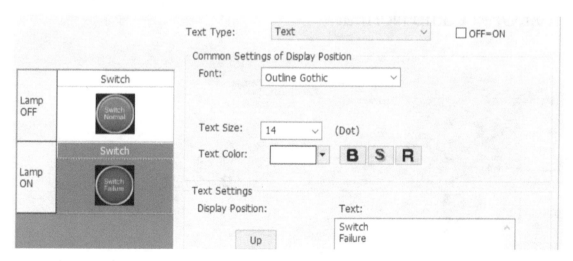

Part location and size
(Displayed in the lower left corner after clicking on a part)
X:32 Y:224 Width:80 Height:80

Button for switching between normal and faulty lamps.
Copy ⇒ Paste switch normal/failure switch button
Basic Settings (Action)
Device: Y0101
Lamp function device: Y0101

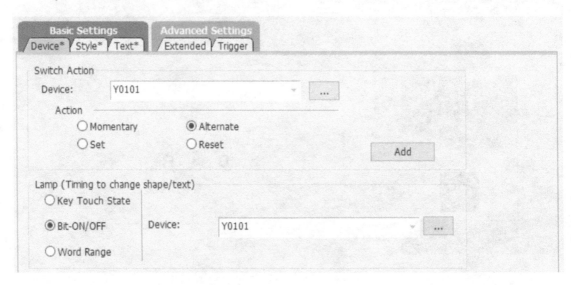

[Basic settings (text)]
Text OFF: Lamp Normal

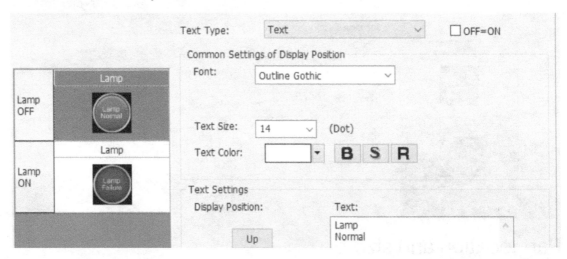

Text ON: Lamp Failure

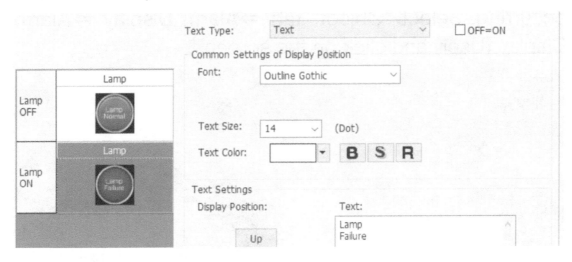

Part location and size
(Displayed in the lower left corner after clicking on a part)
X:32 Y:320 Width:80 Height:80

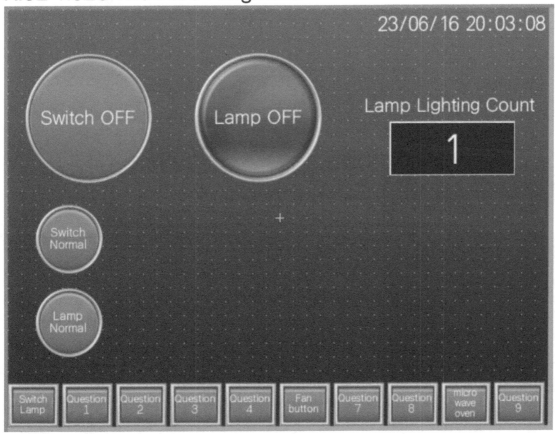

3. Finally, create an alarm display for the alarms that are occurring. Select "Object" tab ⇒Alarm Display ⇒Alarm Display (User) and click on the screen.

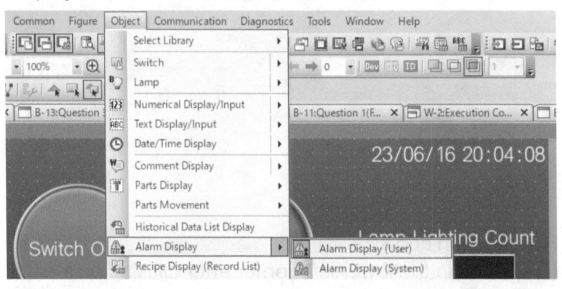

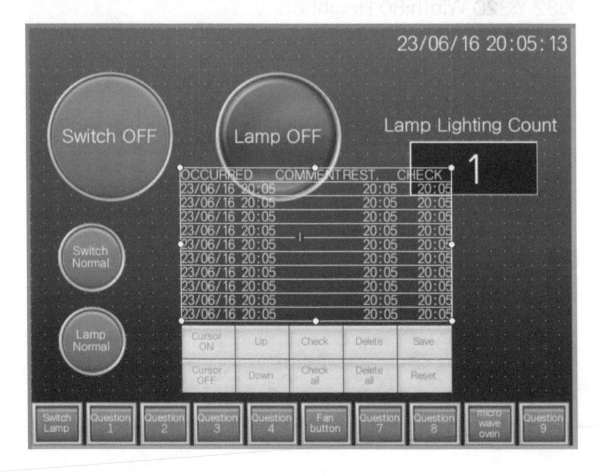

4. delete the "cursor display" and other items under the alarm display area, as they are unnecessary. (They are used in the alarm history.)

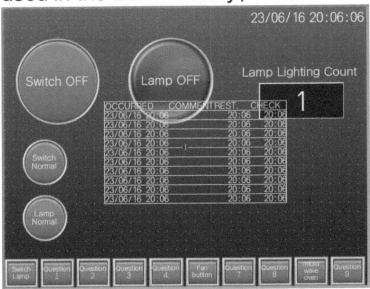

5. Double-click on the alarm display area to configure the alarm display settings. Click "Edit..." under Alarm ID in Basic Settings (Alarm Settings). and press "Yes"(はい).

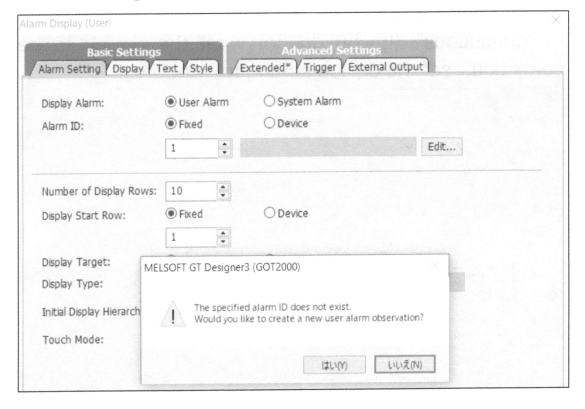

6. The User Alarm Monitoring screen will appear, and in the Basic Settings, set Alarm ID: 1, Alarm Name: Switch/Lamp Alarm Indication, and History Collection Method: Alarms in progress only.

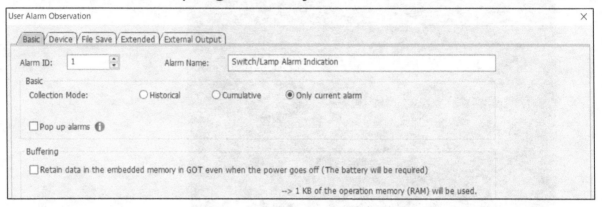

7. Next, set the number of alarm points: 2, comment group: 2, and device 1: Y0100 in the device settings of the user alarm monitoring screen.

Device 2, Y0101, is automatically entered after Y0100 is entered and confirmed because the device setting is set to "continuous". (If you do not want the device to be continuous, set the device setting to "random.)

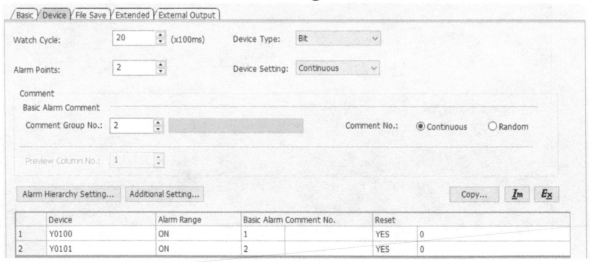

8. Set the alarm comment. Click the "Edit..." button when you click the comment field of basic alarm comment No. 1 in the device settings. Click it, enter "Switch Failure" as the comment, and press "OK".

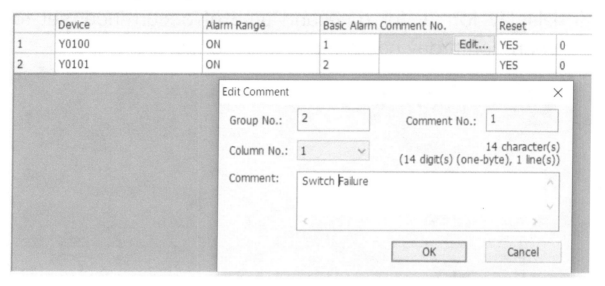

	Device	Alarm Range	Basic Alarm Comment No.		Reset	
1	Y0100	ON	1	Edit...	YES	0
2	Y0101	ON	2		YES	0

Edit Comment ✕

Group No.: 2   Comment No.: 1

Column No.: 1   14 character(s)
(14 digit(s) (one-byte), 1 line(s))

Comment: Switch Failure

OK   Cancel

Follow the same procedure to register "Lamp Failure" as basic alarm comment No. 2, then press "OK" in the lower right corner of the screen to confirm.

	Device	Alarm Range	Basic Alarm Comment No.		Reset	
1	Y0100	ON	1	Switch Failure	YES	0
2	Y0101	ON	2	Lamp Failure	YES	0

9. Return to "Alarm Display (User)," click the Display Items tab in Basic Settings, set the alarm display items as shown below, and press "OK" in the lower right corner.

Display format for date and time of occurrence: set to year/month/day hours/minutes/seconds.

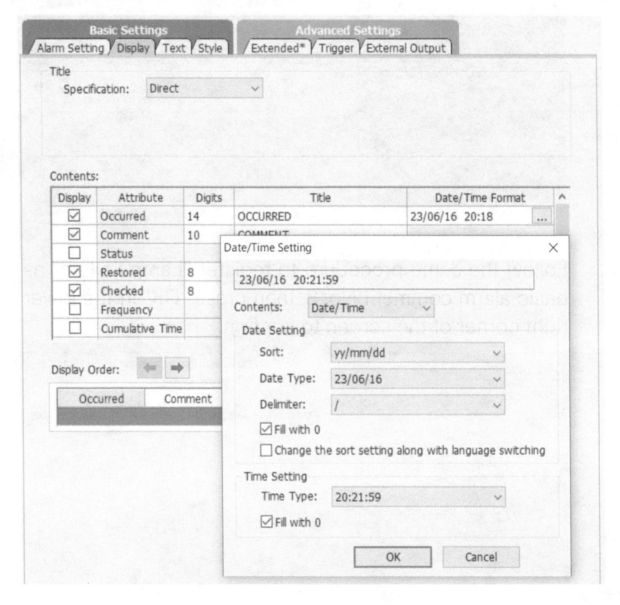

Number of comment digits:20
Comment Title: ALARM NAME
Uncheck display of "Restored" and "Checked".

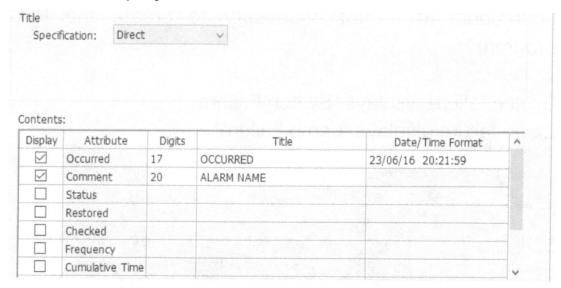

Title
Specification: Direct

Contents:

Display	Attribute	Digits	Title	Date/Time Format
☑	Occurred	17	OCCURRED	23/06/16 20:21:59
☑	Comment	20	ALARM NAME	
☐	Status			
☐	Restored			
☐	Checked			
☐	Frequency			
☐	Cumulative Time			

10. adjust the position of the alarm display.
Part position (displayed at the lower left after clicking on the part) X:160 Y:224

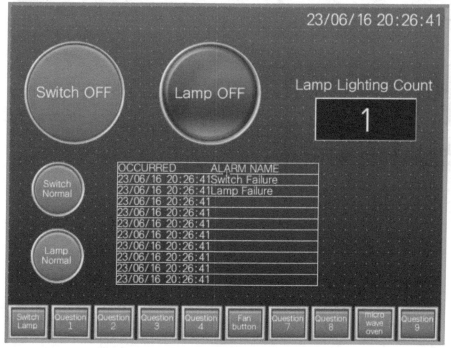

After the screen has been created, start the simulation with the PLC program used in the past simulations, start the simulator for the screen, and confirm the following operations. (It is not necessary to create the PLC program.)

Switch failure displays "Switch Failure"
Lamp failure displays "Lamp Failure"

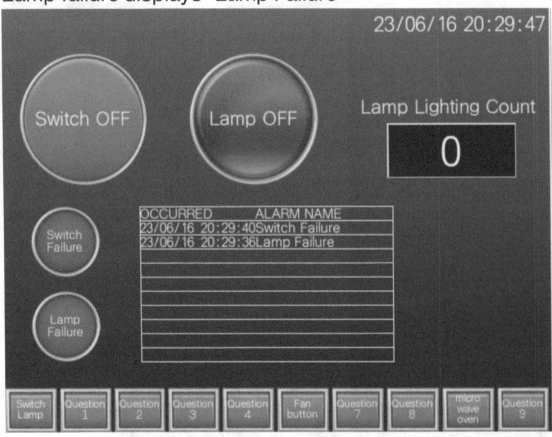

When the lamp is returned to normal, the display disappears.

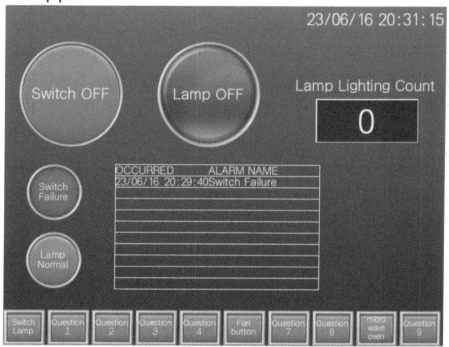

When the switch is returned to normal, the display disappears.

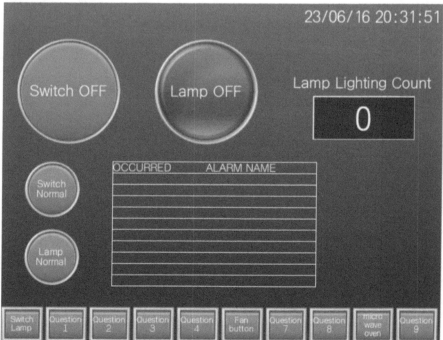

7-4 Alarm History

In the alarm history, you can check the alarms that have occurred in the past in addition to the alarms that are currently occurring when there is an abnormality in the facility. The purpose of this function is to check what alarms have been issued so far. For the creation of alarm history, change the settings for the alarm display on Screen 10 that you just created, and add the "cursor display" and other settings that you deleted in the alarm display.

1. First, change the alarm display settings. Double-click on the alarm display and change the number of displayed lines to 5.

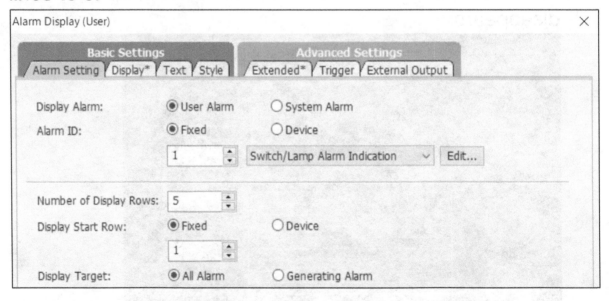

2. Click "Edit..." under Alarm ID in Basic Settings (Alarm Settings) to display the User Alarm Monitoring screen and make the following settings. History collection mode: Historical mode, stored Number: 100 (items)

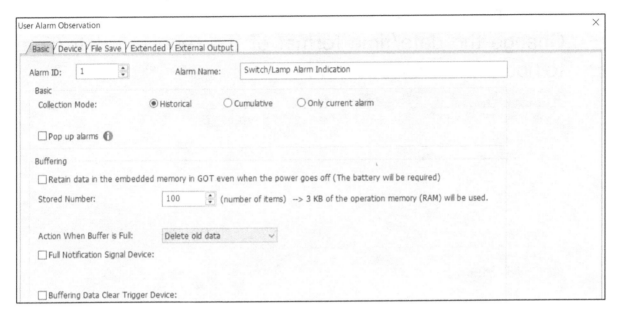

3. Press "OK" in the lower right corner of the user alarm monitoring, then press "Yes"(はい) when the following message is displayed.

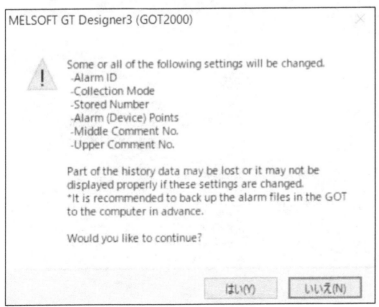

4. Return to "Alarm Display (User)," click the Display Items tab in Basic Settings, set the alarm display items as shown below, and press "OK" in the lower right corner.

Check the check box to display the date and time of restoration.

Change the date/time format of the recovery date/time to hours: minutes: seconds.

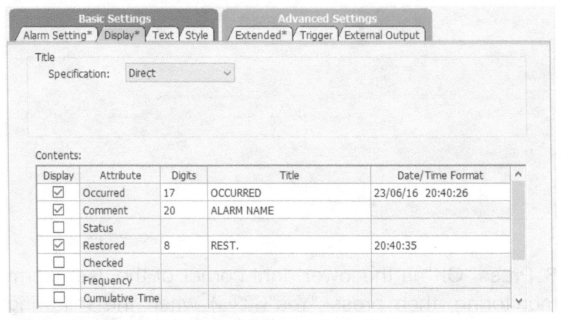

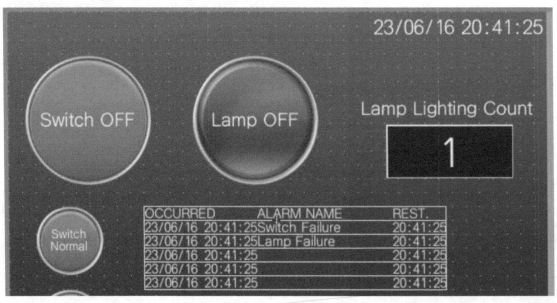

5. add "cursor display" etc. Select "Object" tab ⇒Alarm Display ⇒Alarm Display (User) and click on the screen.

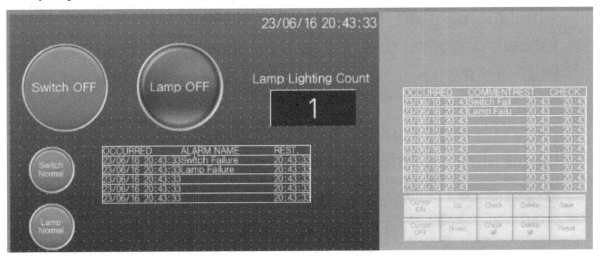

6. delete the alarm display that was just displayed on the screen and move the "cursor display" etc. below the original alarm history display.
Part location and size (Displayed in the lower left corner after clicking on a part) X:160 Y:336

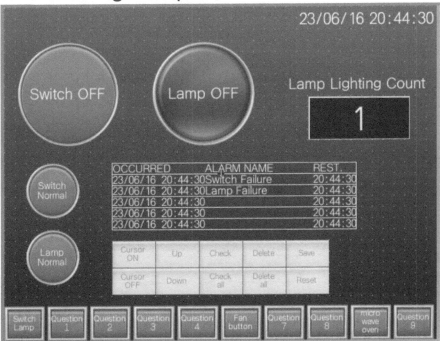

7. After the alarm history has been created, start the simulation with the PLC program used in the past simulations, start the simulator screen, and confirm the following operations. (It is not necessary to create the PLC program.)

Switch failure displays "Switch Failure"
Lamp failure displays "Lamp Failure"

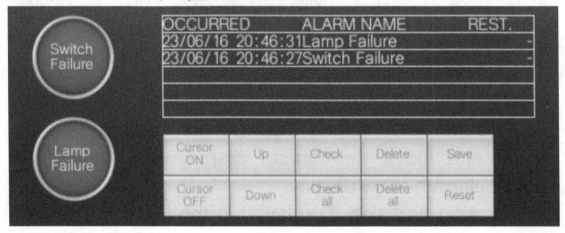

When the switch is returned to normal, the restoration time is displayed.
When the lamp is returned to normal, the recovery time is displayed.

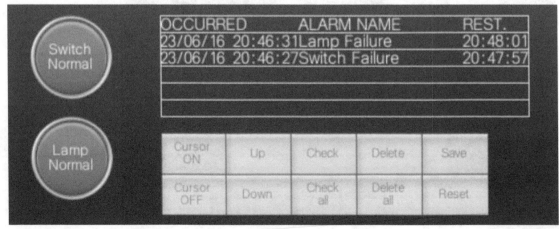

When the switch is set to fail again, a new "Switch Failure" is displayed.

When the lamp is set to "Lamp Failure" again, a new "Lamp Failure" is displayed.

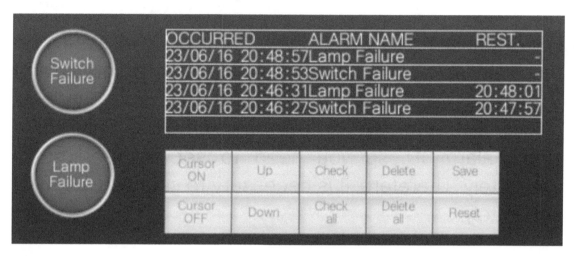

After 6 or more alarms, all alarms can be viewed using "Show Cursor", "Clear Cursor", "Move Up", and "Move Down".

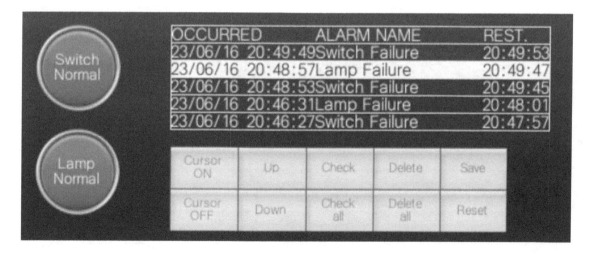

7-5 How to use the display of an occurring alarm and alarm history together.

For example, if you want to view the history of alarms on the main screen by displaying alarms in progress and then pressing the alarm history button, you can only select "history mode" or "alarms in progress only" for a single alarm ID, so you need to set each alarm ID separately. Therefore, it is necessary to set each alarm ID separately.

Example)
Alarm ID for displaying alarms in progress: 10.

User Alarm Observation

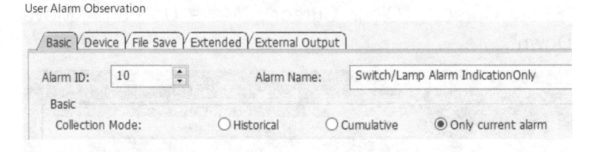

Alarm ID for alarm history display: 11

User Alarm Observation

7-3 to 7-5 Summary

The alarm display can be used to check the alarms that are currently occurring when there is an abnormality in the equipment, for example.

The use is to quickly check only the alarms that are occurring now, for example.

The alarm history allows the user to check past alarms in addition to the current alarm when there is an abnormality at the facility.

The use is to check what alarms have been issued so far.

If you want to display both alarms in progress and alarm history, you need to set separate alarm IDs for each.

Exercise 14 (Conveyor belt sushi order display)

Create a new screen number 60, question 14, and create a screen that displays the order item and order date and time when a person coming to eat presses each order button [action setting: bit set], and that order item and order date and time disappear when the kitchen staff presses each serving completion button [action setting: bit reset]. Use X0050-X0053 as the device for each order button and 2 as the alarm ID. (Please also add a button to move between screens 10-60)

Order material and order date and time are displayed when each order button is pressed.

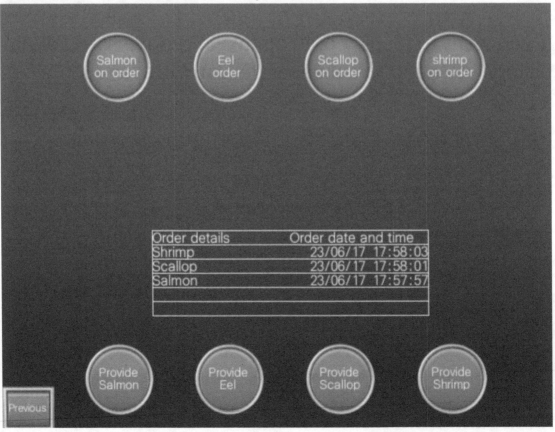

When you click the Offer Complete button for the item being ordered, the ordered item and the date and time of the order disappear.

Add a Go to Screen 60 (Problem 14) button on Screen 10 and a Go to Screen 10 button on Screen 60, respectively.

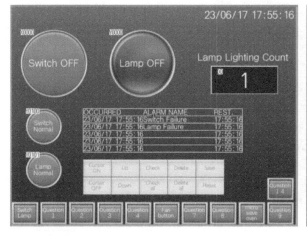 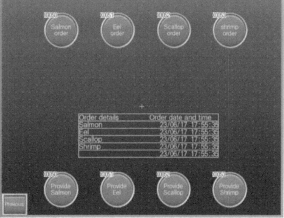

Exercise 14 Solution

1. after creating a new screen number 60, question 14, first create each order button.

Salmon Order Button
"Object" tab => Switch ⇒ Bit Switch

Basic Settings (Action)
Device: X0050
Action: Set
Lamp function: Bit ON/OFF
Lamp function device: X0050

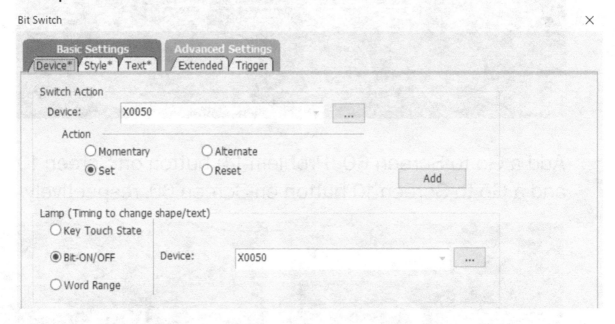

*After that, you arefree to set up as you wish.

(Reference setting)
Basic Settings (Style)
OFF shape color: blue
OFF shape: FA Control Switch Figure A1 1 SW_01_0_B

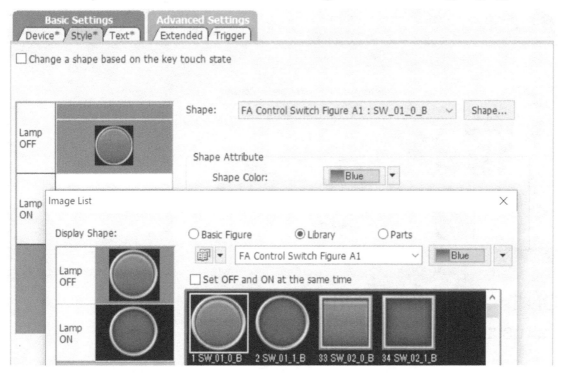

ON shape color: green
ON shape: FA Control Switch Figure A1 8 SW_01_1_G

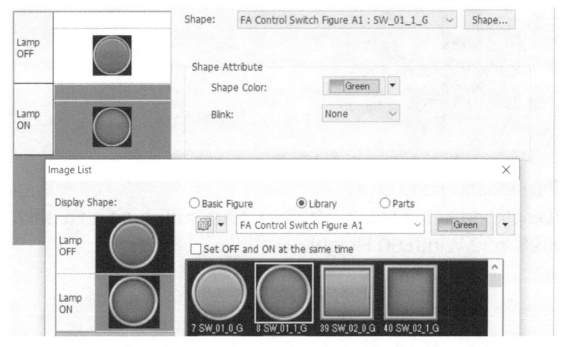

[Basic settings (text)]
Text type OFF=ON check: OFF
OFF text: Salmon order
Text size: 14

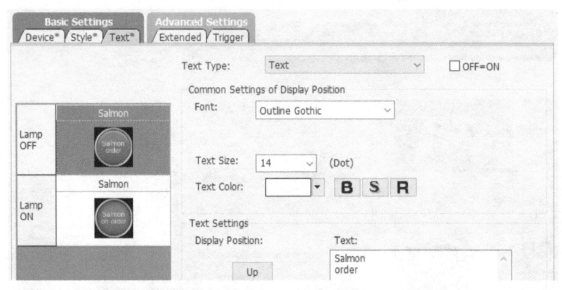

ON Text: Salmon on order
Text size: 14

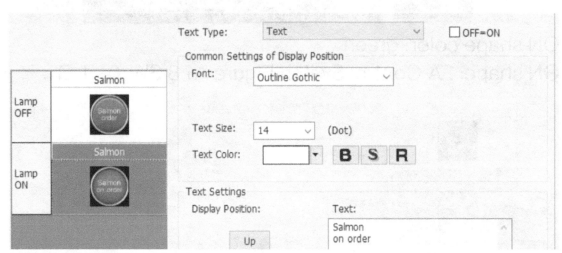

Part location and size
(Displayed in the lower left corner after clicking on a part)
X:96 Y:32 Width:80 Height:80

Eel/Scallop/Shrimp order button
Create a continuous copy of the Salmon Order button.
Total count after copy   X:4 Y:1
Interval : Interval   Distance X: 50
Increment target: device No.
Increment setting: All、1

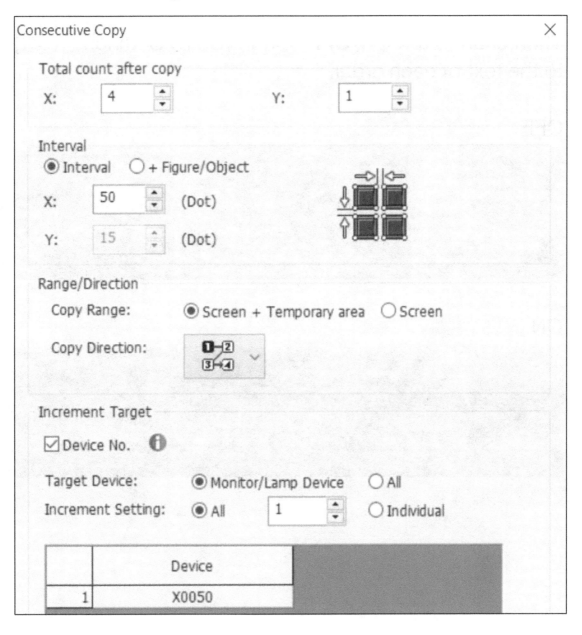

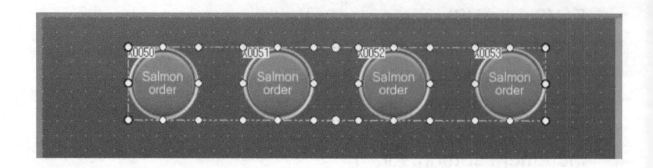

Then change the ON/OFF text in [Basic Settings (Text)] to the text of each order.

OFF

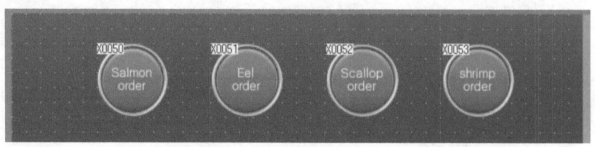

ON

334

2 Next, we will create a button for the completion of each offer.

Provide Salmon Button
"Object" tab => Switch ⇒ Bit Switch

Basic Settings (Action)
Device: X0050
Action: Reset.

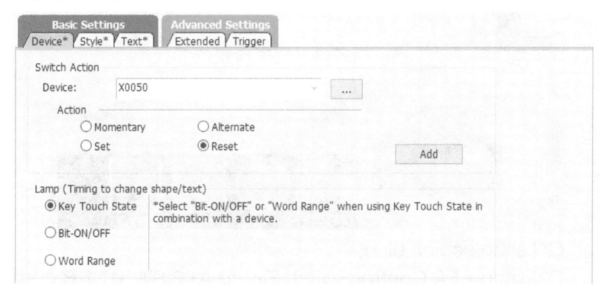

*After that, you are free to set up as you wish.

(Reference setting)
Basic Settings (Style)
OFF shape color: blue
OFF shape: FA Control Switch Figure A1 1 SW_01_0_B

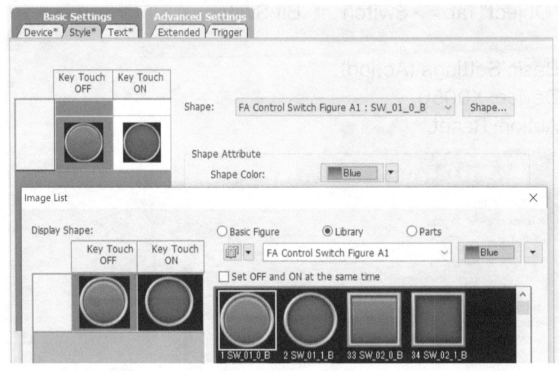

ON shape color: blue

ON shape: FA Control Switch Figure A1 2 SW_01_1_B

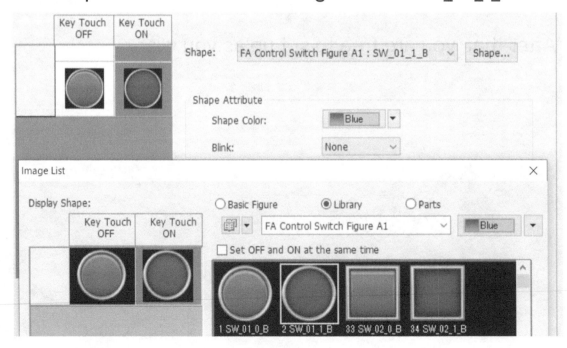

[Basic settings (text)]
ON/OFF text: Provide Salmon
Text Size: 14

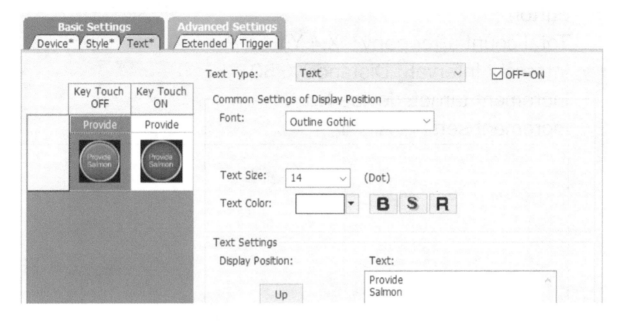

Part location and size
(Displayed in the lower left corner after clicking on a part)
X:96 Y:384 Width:80 Height:80

Provide Eel/Scallop/Shrimp Button

Create a continuous copy of the Salmon Offer Complete button.
Total count after copy   X:4 Y:1
Interval : Interval   Distance X: 50
Increment target: device No.
Increment setting: All、1

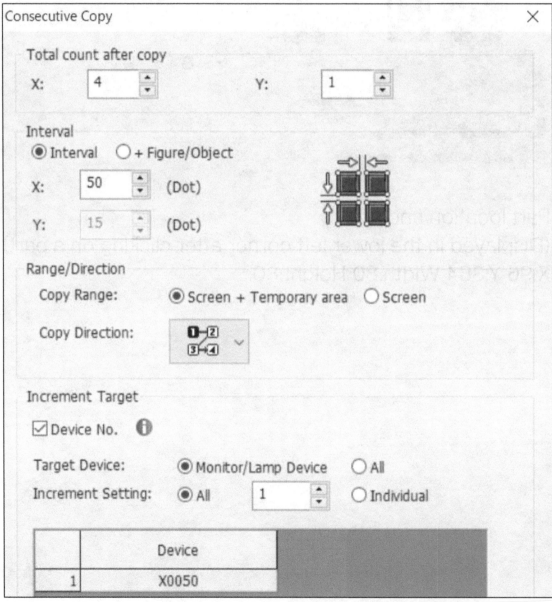

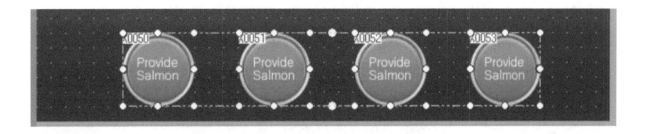

Then change the ON/OFF text in [Basic Settings (Text)] to each offering completion story.

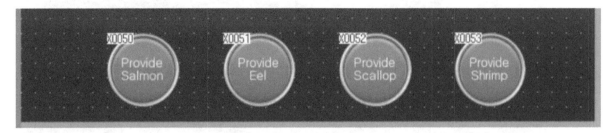

3. Create a place to display the order story and the date and time of the order. Select "Object" tab ⇒ Alarm Display ⇒Alarm Display (User), then click on the screen.

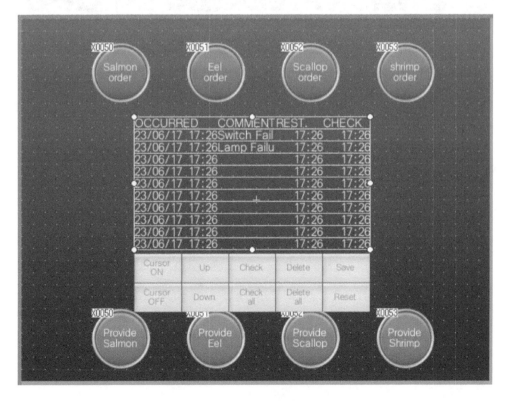

4. remove the "cursor display" and other items under the alarm display area, as they are unnecessary.

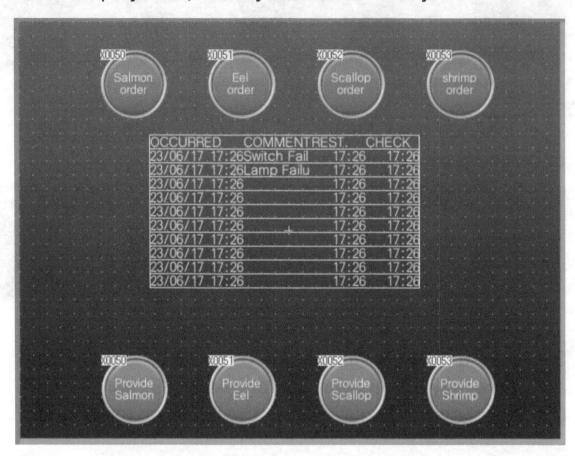

5. Double-click on the alarm display area to configure the alarm display settings. After changing the number of display digits to 5 and the alarm ID to 2 in the Basic Settings (Alarm Settings), click "Edit..." and a message will appearasking if you want to create a new user alarm monitoring. Click "Edit..." and then click "Yes"(はい) to create a new user alarm monitoring?

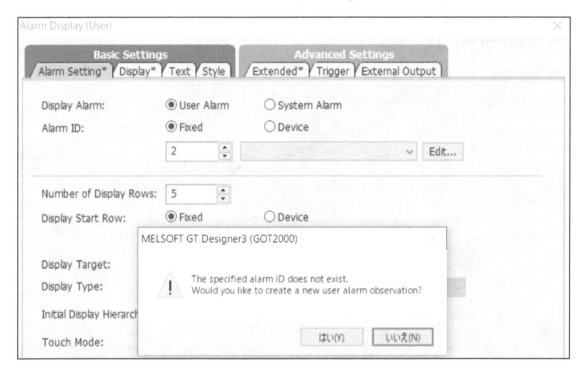

6. User Alarm Monitoring will be displayed. In the Basic Settings, set Alarm Name: Conveyorbelt sushi order display, and History Collection Method: Only current alarm.

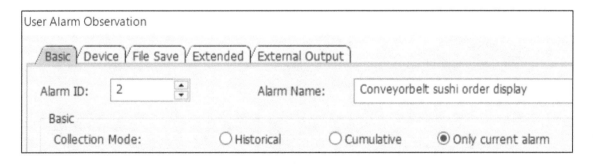

7. Next, in the device settings for user alarm monitoring, set the number of alarm points: 4, comment group: 3, and device 1: X0050.

Devices 2 to 4, X0051 to X0053, are automatically entered after X0050 is entered and confirmed because the device setting is set to "Continuous". (If you do not want the devices to be continuous, set the device setting to "Random".)

User Alarm Observation

| Basic | Device | File Save | Extended | External Output |

| Watch Cycle: | 20 | (x100ms) | Device Type: | Bit |
| Alarm Points: | 4 | | Device Setting: | Continuous |

Comment
Basic Alarm Comment

| Comment Group No.: | 3 | | Comment No.: | ⦿ Continuous |

Preview Column No.: 1

Alarm Hierarchy Setting...   Additional Setting...   Co

	Device	Alarm Range	Basic Alarm Comment No.		Reset	
1	X0050	ON	1		YES	0
2	X0051	ON	2		YES	0
3	X0052	ON	3		YES	0
4	X0053	ON	4		YES	0

8. Set the alarm comment. In the device settings, clickon the comment field of the basic alarm comment No. 1 to display the "Edit..." button, click on it, enter "Salmon" for the comment, and press "OK".

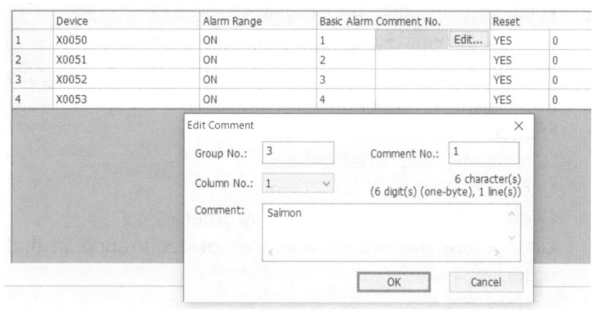

	Device	Alarm Range	Basic Alarm Comment No.			Reset	
1	X0050	ON	1		Edit...	YES	0
2	X0051	ON	2			YES	0
3	X0052	ON	3			YES	0
4	X0053	ON	4			YES	0

Edit Comment ✕

Group No.: 3   Comment No.: 1

Column No.: 1   6 character(s)
(6 digit(s) (one-byte), 1 line(s))

Comment: Salmon

OK   Cancel

Follow the same procedure to register "Eel" for basic alarm comment No. 2, "Scallop" for No. 3, and "Shrimp" for No. 4, then press "OK" in the lower right corner of the screen to confirm.

	Device	Alarm Range	Basic Alarm Comment No.		Reset	
1	X0050	ON	1	Salmon	YES	0
2	X0051	ON	2	Eel	YES	0
3	X0052	ON	3	Scallop	YES	0
4	X0053	ON	4	Shrimp	YES	0

9. Return to "Alarm Display (User)," click the Display Items tab in Basic Settings, set the alarm display items as shown below, and press "OK" in the lower right corner.

Number of "Occurred "digits:20.
"Occurred" title: Order date and time.
Display format for "Occurred": set to year/month/day hours/minutes/seconds.
Number of "Comment "digits:20.
"Comment" title: Order details
Uncheck "Restored" and "Checked".
Click on the arrows in the display order.
Change comment, date, and time of occurrence, in that order.

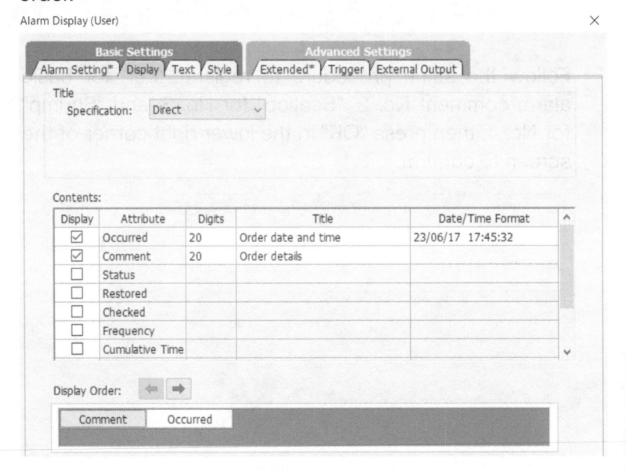

10. adjust the position of the order display.
Part position (displayed at the lower left after clicking on the part)
X:176 Y:256

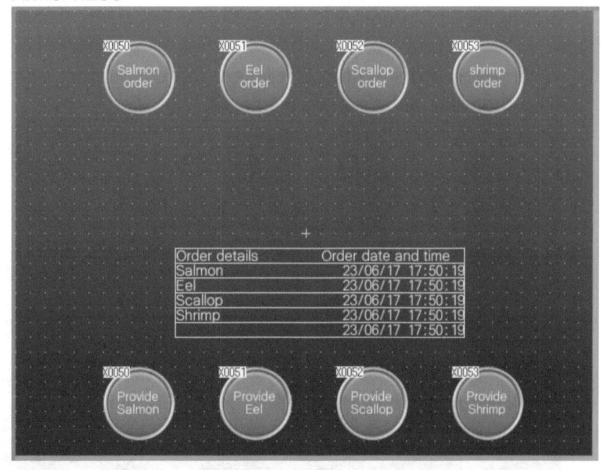

11. finally, add move buttons to screen 10 and screen 60.

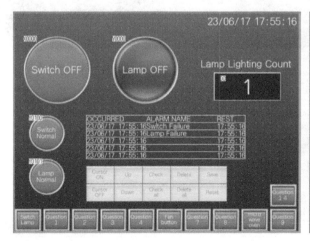

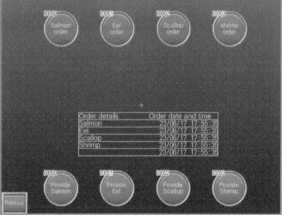

12. Now that you have finished creating the screen, start the simulation with the PLC program used in the past simulations, start the simulator for the screen, and confirm the following operations. (It is not necessary to create the PLC program.)

When each order button is clicked, the order story and the order date and time are displayed.

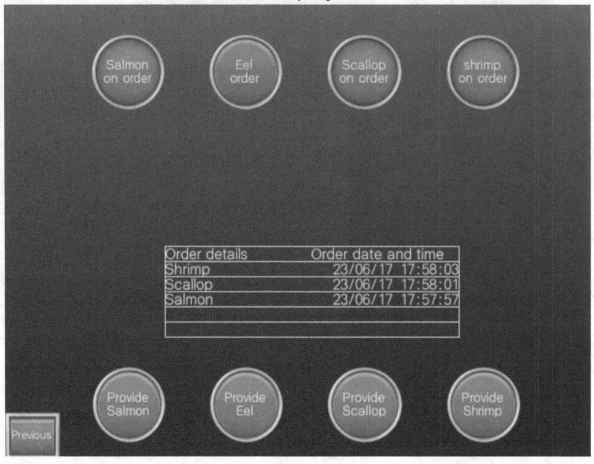

When you click the "Provide" button for the item you are ordering, the item you ordered, and the date and time of the order will disappear.

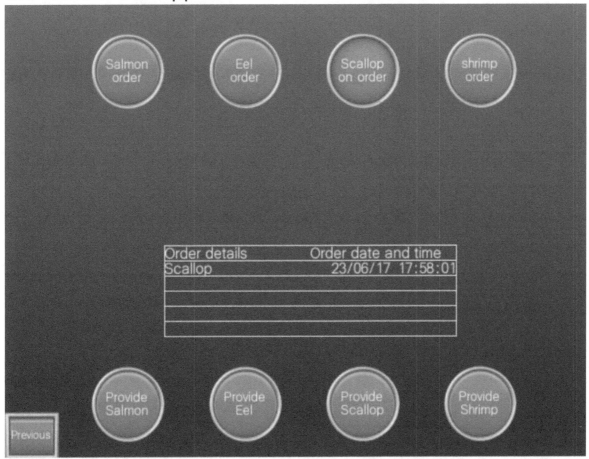

Exercise 15 (Conveyorbelt sushi order history display)

Please make sure that the order history is also kept in the display of screen number 60, question 14.

Order material and order date and time are displayed when each order button is pressed.

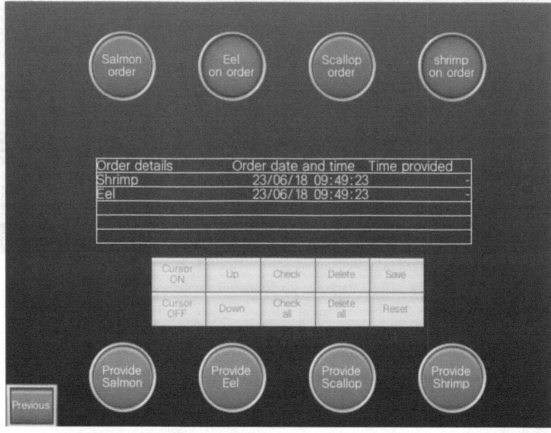

When you press the "Finished providing" button for the item you are ordering, the finished providing date and time are entered, and the ordered item and order date and time remain in the history.

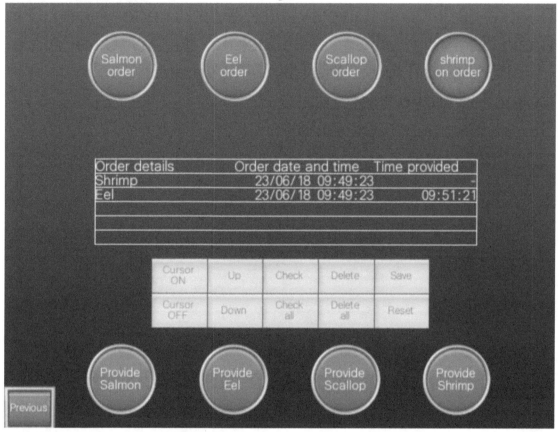

Exercise 15 Solution

1. First, change the settings of the order display. Double-click Order Display, click "Edit..." under Alarm ID in Basic Settings (Alarm Settings) to display User Alarm Monitoring, and make the following settings.
Collection Mode: Historical
Stored Number: 100 (items)

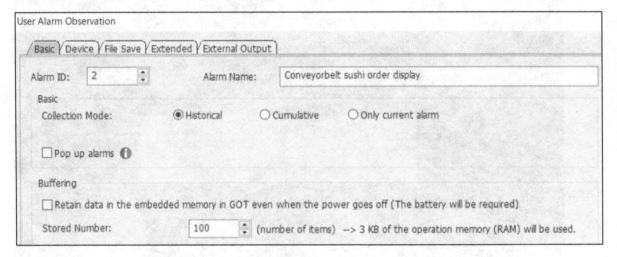

2. Press "OK" in the lower right corner of the user alarm monitoring, then press "Yes"(はい) when the following message is displayed.

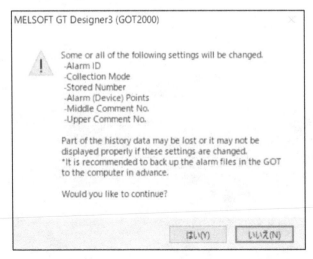

3. Return to "Alarm Display (User)," click the Display Items tab in Basic Settings, set the alarm display items as shown below, and press "OK" in the lower right corner.

Restored : Check.

Digits: 15

Title : Time provided

Date/Time Format : hour: minute: second

Contents:

Display	Attribute	Digits	Title	Date/Time Format	^
☑	Occurred	20	Order date and time	23/06/18 09:33:49	
☑	Comment	20	Order details		
☐	Status				
☑	Restored	15	Time provided	09:38:57	
☐	Checked				
☐	Frequency				
☐	Cumulative Time				∨

Display Order: ⬅ ➡

Comment	Occurred	Restored	

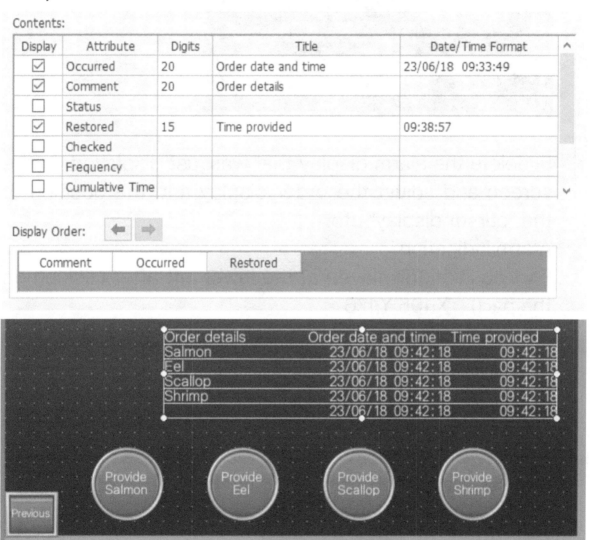

4. add "cursor display" etc. Select "Object" tab ⇒Alarm Display ⇒Alarm Display (User) and click on the screen.

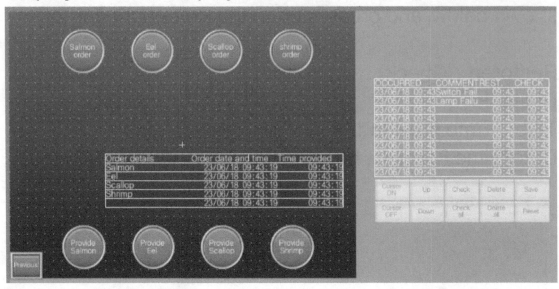

5. delete the alarm display that was just displayed on the screen and adjust the order display and the position of the "cursor display" etc.
Alarm indication
Part position (displayed at the lower left after clicking on the part)   X:105 Y:176
Cursor display, etc.
Part position (displayed at the lower left after clicking on the part)   X:170 Y:288

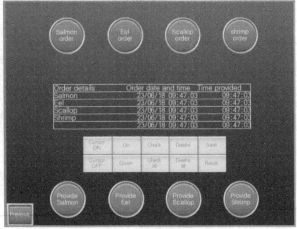

6. Now that the order display and order history have been created, start the simulation with the PLC program used in the past simulations, and start the simulator to confirm the following operations on the screen as well. (It is not necessary to create the PLC program.)

When each order button is clicked, the order information and the order date and time are displayed.

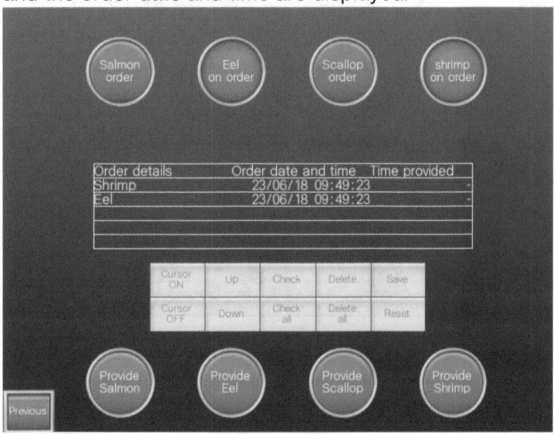

When you press the "Finished providing" button for the item you are ordering, the time when the order is complete is entered, and the item you ordered, and the date and time of the order remain in the history.

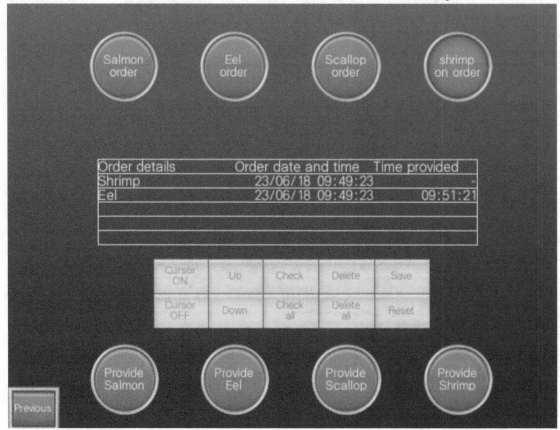

After displaying 6 or more orders and order history, all orders and order history can be viewed using "Cursor ON", "Cursor OFF", "Up", and "Down".

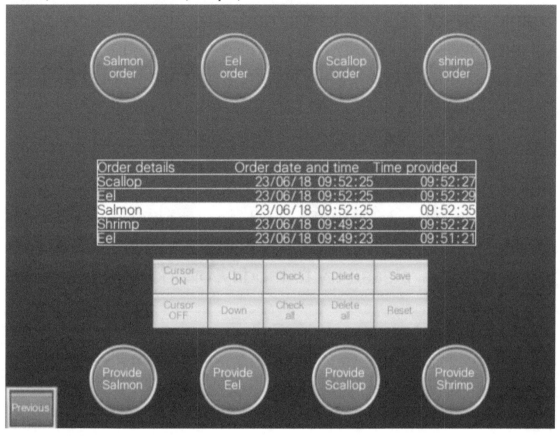

At the end.

Thank you for reading to the end. Creating a touch panel screen involves creating it, running it, repeating trial, and error, and finally completing what you want to create is an indescribable feeling. And repetition is the fastest way to improve skills. By repeating the exercises in this book, you will acquire the basic skills, but after that, we urge you to think about your own assignments and the screens you would like to create and try creating your own screens. And if the company or factory you work for uses touch screens, please try to modify, or add screens.

As I mentioned at the beginning,
I want to learn about touch panels for my equipment because I'm going to use them at work!"
There is no book that goes into detail about equipment touch panels."
I've read books on touchscreens, but I don't feel like I can create a screen just from classroom learning."
I'm studying throughe-learning, but the content is so boring that I've given up..."

We hope that through this book, these people will find creating touch-panel screens a little more enjoyable.
We look forward to your success in the various industries in which you have read this book!

Made in United States
Cleveland, OH
08 May 2025

16759765R00197